Digital Radio DAB+

Marian Oziewicz

Digital Radio DAB+

Broadcasting Multimedia System

 Springer

Marian Oziewicz
National Institute of Telecommunications
Wroclaw, Poland

ISBN 978-3-030-66480-0   ISBN 978-3-030-66478-7   (eBook)
https://doi.org/10.1007/978-3-030-66478-7

This Springer imprint is published by the registered company Springer Nature Switzerland AG
The registered company address is: Gewerbestrasse 11, 6330 Cham, Switzerland

The author of the present book was a finalist at the International Mathematical Olympiad. He graduated from the Faculty of Electronics, Technical University of Gdańsk, Poland. Probationary period at the Technical University was paused to study quantum field theory in the Department of Theoretical Physics, University of Wroclaw. Introduction of the martial law in Poland interrupted work on completing a doctorate in theory of elementary particles.

The conditions of martial law forced him to change jobs and turn back to the problems of telecommunications. The opportunity to become familiar with system of the digital radio DAB in various stages of its development appeared in the leading European research centers. In 1992, he was invited to the Institute IRT in Munich, a member of the EU project Eureka 147 DAB, where he was introduced to the core concepts of signal processing in digital transmitter and receiver. This took place still before the final determination and standardization of the system. Presentation of the system was carried out using a prototype of the variant parameters.

Further studies of the DAB system were conducted in the framework of the EU grant 'visiting scientist' in scientific and research multimedia laboratory in Hildesheim in Germany.
His Ph.D. in the field of telecommunications was completed at the Department of Radioelectronics, Warsaw University of Technology.

Preface

The work on digital radio in Europe started in the framework of the European Union (UE) project Eureka 147 DAB in 1987. The aim was to develop a fully digital radio system for fixed, portable, and mobile reception of multimedia. The work of members of the project resulted in the European Norm ETSI EN 300 401 highlighted as the EU digital radio standard DAB – Digital Audio Broadcasting. Further development of the system was conducted by the different working groups of international experts under the auspices of such organizations as EBU (European Broadcasting Union), focusing representatives of public broadcasters, and ETSI (European Telecommunication Standards Institute), responsible for the European standards in the field of telecommunications, further coordinating within JTC (Joint Technical Committee) and including later also CENELEC (Comite European de Normalisation ELECtrotechnique) responsible for standardizing radio.

The essential element of digital radio is the audio encoder, which reduces the number of audio parameters. In DAB, this is the MUSICAM encoder. However, with the development of audio compression techniques, a more efficient HE-AACv2 type encoder was introduced into the system along with a reduction of some other options.

This improved system is referred to as DAB+ radio.

In this book, we deal with the basic mechanisms of digital radio (technical details can be found in the System Specifications) – therefore, apart from the parts describing the audio encoders, the acronyms DAB and DAB+ can be treated equally.

The concepts and methods developed in the framework of the Digital Radio project in Europe were next used in digital television and the latest generations of mobile phones.

Apart from work on the technical core of the system in developed EU countries, action has been taken on developing multimedia applications of digital broadcasting. It is not only an extension of information accompanying program, but also audio independent so-called value-added services including messages of social relevance, such as a warning system for the civilian population, traffic and transport messages, information on local tourist services, and parking. The advertising information should be significantly expanded, taking advantage of the graphic exposure options.

The role of the radio – due to its universal access and the forms of communication – is not so completely absorbing as television or telephone. Radio is indispensable as a means of information of the drivers. Similarly, in situations of risk, flood, or fire, where the cellular communications are blocked due to an overload capacity of the cells by the number of entries, the optimal source of information is the radio.

Applications of digital radio will depend on the creativity of its operators: new technology can be used passively as replicate of the analog radio transmission limited to the verbal and musical programs, or actively exploited developing the possibilities of parallel services based on multimedia data.

Currently, description of the DAB system is contained in more than 80 specifications. Working on a uniform, comprehensive description of the entirety, one should select its basic elements not found in analog radio system. The focus is therefore on description of the mechanisms extending the radio transmission on the multimedia content. The standards and specifications describe the exact organization of protocols, interfaces, and components of the physical and higher layers of the system, but do not devote space to explaining why and how such procedures are recommended. The book takes appropriate explanation on the assumed level and shows the current state of the general structure of the DAB and DAB+ systems.

Wroclaw, Poland   Marian Oziewicz

Admission

The motivations for initiating work on the system Digital Audio Broadcasting (DAB), and its following version DAB+ with a new audio encoder, on the technical side are:

- Inclusion of the radio in the digital teletransmission world allowing for signal processing, extension of audio signals to multimedia messages, and archiving programs with digital methods
- Extending the coverage area of the radio program of single transmitter to broadcasting area of many transmitters connected in a Single Frequency Network (SFN)
- Optimal use of scarce goods such as radio frequency spectrum, by replacement of the channel bandwidth allocation by the grants of digital throughput adapted to the current needs of programs under a common block of spectrum

The DAB+ is a digital broadcasting system designed for reliable use also in mobile reception. The information transferred in this system can be of different kinds. The sound track in DAB radio, justifying its name, stems from organic incorporation of the audio encoder in the transmitter and decoder in the receiver. But it can also be transparent for other digital data with capacity allowed by its throughput.

In determining the relationship of DAB+ to earlier systems, it should be noted that digital audio broadcasting is a new value compared to the existing AM, FM, or FM stereo radios.

The implementation of DAB+ demands new transmitters and receivers: at homes, in cars, and, in the future, in cellular phones.

Services offered by this system can be extended for "information on demand" using text, graphics, images, and video. Tests with broadcasting a television program via the DAB system revealed that mobile reliability and stability of its reception is incomparably better than in analog systems under the same conditions.

DAB system helped optimize the use of frequencies for planning both local radio and networks of transmitters covering larger areas of the country.

   As for any digital system, the transmission of DAB+ radio programs can be described in the most adequate way by the OSI system of layers: the physical layer, the network layer, the transport layer, management, and the presentation layer.

   The physical layer of the radio describes the functional blocks of the system.

   The network layer describes how to format the data in the logical frames of DAB+.

   For the skillful use of the DAB+ opportunities, the transport layer describing the organization of the output signal of encoders of various applications is equally important.

   The bit rate of individual applications is controlled by a management system organized and steered by the Fast Information Channel.

   These layers, as in the description of the other digital tele-transmission systems, are interdependent and interconnected.

   Our aim is to introduce to the indicated issues.

   The encoder/decoder systems in DAB and DAB+ are different, but the principles of organization of both systems are alike, although not all elements apply in both cases. Because of this, after the part describing encoders, the acronyms DAB and DAB+ are used equally, unless otherwise stated.

Contents

List of Acronyms

AAC	Advanced Audio Coding
ACS	Access Control System
AIC	Auxiliary Information Channel
AIFF	Audio Interchange File Format
ALC	Asynchronous Layered Coding
APNG	Animated Portable Network Graphics
ASCII	American Standard Code for Information Interchange
ASF	Advanced Streaming Format
ASCTy	Audio Service Component Type
ASu	Announcement Support (flags)
ASw	Announcement Switching (flags)
ATRAC	Adaptive Transform Acoustic Coding
AU	Access Unit
AV	Audio-Visual
A/V	Audio/Video
BER	Bit Error Ratio
BIFS	Binary Format for Scene
BNS	Broadcast Network Server
Bslbf	bit string, left bit first
BWS	Broadcast Web Site
CA	Conditional Access
CAT	Conditional Access Table
CCA	Component Conditional Access
CBMS	Convergence of Broadcast and Mobile Services
CEI	Change Event Indication
CGI	Common Gateway Interface
CI	Content Indicator
CIF	Common Interleave Frame
ClusterId	Cluster Identifier
CMD	Command code identifying the message category in STI interface
C/N	Carrier-to-Noise (Ratio)

COFDM	Coded Orthogonal Frequency Division Multiplex
CRC	Cyclic Redundancy Check
CU	Capacity unit
CW	Control Word
DAB	Digital Audio Broadcasting
DAB+	DAB with AAC codec
DFT	Digital Fourier Transform
DG	Data Group
DGCA	Data Group Conditional Access
DGI	DAB Gateway Interface
DLS	Dynamic Label Segment
DMB	Digital Multimedia Broadcasting
D-PSK	Differential Phase Shift Keying
DRM	Digital Radio Mondiale
DSCTy	Data Service Component Type
DSP	Digital Signal Processor
DVB	Digital Video Broadcasting
DVB-H	DVB-Handheld
DVB-T	DVB-Terrestrial
EBU-UER	European Broadcasting Union - Union Europeenne de RadioTV
ECC	Extended Country Code
ECM	Entitlement Checking Message
ED	Energy Dispersal
EMM	Entitlement Management Message
EOF	End of Frame
EOH	End of Header
EPG	Electronic Programme Guide
EPID	Ensemble Provider Identifier
EPM	Enhanced Packet Mode
ES	Elementary Stream
ESG	Electronic Service Guide
ESM	Enhanced Stream Mode
ETI	Ensemble Transport Interface
ETI(NA)	ETI Network Adaptation
ETI(NI)	ETI Network Independent
ETS	European Telecommunication Standard
ETSI	European Telecommunications Standards Institute
FC	Frame Characterization
FEC	Forward Error Correction
FFT	Fast Fourier Transform
FIB	Fast Information Block
FIC	Fast Information Channel
FIDC	Fast Information Data Channel
FIDCCA	Fast Information Data Channel Conditional Access

FIG	Fast Information Group
FLUTE	File de Livery over Unidirectional Transport
F– PAD	Fixed Programme Associated Data
FRPD	Frame Padding
GCA	Group Customer Address
GIF	Graphic Interchange Format
GNSS	Global Network Satellite System
HDTV	High Definition Television
HE AAC	High-Efficiency Advanced Audio Coding
HF	High Frequency
HTML	Hypertext Markup Language
HTTP	Hyper Text Transfer Protocol
I signal	In-phase
ICI	Inter-Carrier-Interference
IEC	International Electrotechnical Committee
IETF	Internet Engineering Task Force
IFFT	Inverse Fast Fourier Transform
IK	Issuer Key
IM	Initialization Modifier
INS	Interaction Network Server
IOD	Initial Object Descriptor
IP	Internet Protocol
IPDC	IP Data Casting
IPDC	International Programme for the Development of Communication
ISI	Inter-Symbol Interference
ISO	International Standardization Organization
ITU	International Telecommunications Union
IW	Initialization Word
JFIF	JPEG File Interchange Format
JPEG	Joint Photographic Expert Group
JTC1	Joint Technical Committee on Information Technology
LCT	Layered Coding Transport
LTO	Local Time Offset
MainId	Main Identifier
MBMS	Multimedia Broadcast/Multicast Service
MCI	Multiplex Configuration Information
MDP	Multipath Delay Profile
MEMO	Multimedia Environment for Mobiles
MFN	Multiple Frequency Network
MHEG	Multimedia and Hypermedia Information Coding Expert Group
MIME	Multipurpose Internet Mail Extensions
MJD	Modified Julian Date
MNS	Multiple Network Server
MOT	Multimedia Object Transfer protocol
MOT BWS	MOT Broadcast Website

MPE	Multi-Protocol Encapsulation
MPEG	Moving Pictures Expert Group
MPEG-2 TS	MPEG-2 Transport Stream
MP2	Multimedia Protocol, Layer 2
MP3	Multimedia Protocol, Layer 3
MSC	Main Service Channel
MST	Main Stream Characterization; Main Stream data
MUSICAM	Masking Pattern Adapted Universal Sub-band Integrated Coding and Multiplexing
NIT	Network Information Table
OD	Object Descriptor
OFDM	Orthogonal Frequency Division Multiplex
O&M	Operation and Management System
PACT	President's Advisory Committee on Future Technology
PAD	Programme Associated Data
PAT	Programme Association Table
PCM	Pulse Code Modulation
PCR	Programme Clock Reference
PDA	Personal Digital Assistant
PDK	Programme Distribution Key
PES	Packetized Elementary Stream
PI	Punctured Index
PID	Packet Identifier
PLI	Parameter Length Indicator
PMT	Programme Map Table
PNG	Portable Network Graphics
PPID	Programme Provider Identifier
PPUA	Programme Provider Unique Address
PRBS	Pseudo Random Binary Sequence
PS	Parametric Stereo
PSI	Programme-Specific Information
PSNR	Peak Signal-to-Noise Ratio
PTy	Programme Type
Q signal	Quadrature signal
RDS	Radio Data System
RFC	Request for Comments
Rfa	Reserved for future addition
Rfu	Reserved for future use
RS	Reed-Solomon (code)
SA	Service Address
SAT	Sub-channel Assignment Table
SBR	Spectral Band Replication
SCCA	Service Component Conditional Access
SCTy	Service Component Type
SDP	Session Description Protocol

SDT	Service Description Table
SI	Service Information
Sid	Service Identifier
SIV	Service Information Version
SK	Service Key
SL	Synchronization Layer
SPID	Service Provider Identifier
SPS	Service Provider Server
SRTP	Secure Real-Time Transport Protocol
SSCTy	Specific Service Component Type
STC	Stream Characterization
STI	Service Transport Interface
STI-C	STI-Control Part
STI-C(LI)	STI-C Logical Interface
STI-D	STI-Data Part
STI-D(LI)	STI-D Logical Interface
Subsid	Sub-Identifier
Subsid	Sub-channel Identifier
TC	Technical Committee
TCP	Transmission Control Protocol
TCId	Transmission Component Type Identifier
TDC	Transparent Data Channel
TDT	Time and Date Table
TII	Transmitter Identification Information
TIST	Time Stamp
TN	Top News
TMId	Transport Mode Identity/Transport Mechanism Identifier
TPEG	Transport Protocol Experts Group
TS	Transport Stream
TSDT	Transport Stream Description Table
UA	User Application
UA	Unique Address
UDP	User Datagram Protocol
Umsbf	Unsigned Integer, Most Significant Bit First
UMTS	Universal Mobile Telecommunications System
URI	Uniform Resource Identifier
URL	Uniform Resource Locator
URN	Uniform Resource Name
UTC	Universal Time Coordinated
VBR	Variable Bit Rate
XML	eXtensible Markup Language
X-PAD	eXtended Programme Associated Data

Internet Addresses Associated with the DAB

events@worlddab.org	– Information on conferences organized under the auspices of WorldDAB
http://www.worlddmb.org/	– Address of the Digital Mobile Broadcasting association
http://docbox.etsi.org/reference	– Documents of the ETSI organization
http://tech.ebu.ch	– Information about the EBU's activities, including publications on DAB, DAB+, DMB
http://www.tpeg.org/	– Information about the organization TPEG
http://www.adept.eu.com/	– Information about the association ADEPT

Chapter 1
Introduction

For some time, the radio over Internet is a fact.

Internet via radio is the ability opened by the digital multimedia radio DAB (Digital Audio Broadcasting) and its newer version DAB+ [1, 2].

This is a new perspective of radio. A radio enriched – beyond traditional programs of audio and music – by the resources of possible added information. DAB is a multimedia radio allowing transmission of audio, text, graphics, photographs, maps, charts, and video.

The position of the modern radio in the digital communication world is due to its basic features:

- Digital transmission techniques
- Unilateral simultaneous transmission (broadcasting) for large social groups
- Capabilities of multimedia transport

Digitalization of radio – a necessary condition for the transmission of multimedia data – required new concepts and techniques developed in the framework of the EU Project Eureka 147 DAB and further improved within radio institutions under the WorldDAB organization.

1.1 Why Digital Radio

Capabilities of the analog signal processing are limited and gradually exhausting. Digital signals can be precisely recorded, long-lasting after archiving, and processed by methods not possible in the analog technique. The concept of compact discs (Philips) initiated digital signal processing in the radio studio. To achieve full benefits of radio signal, it has become natural to undertake the task of digitization also the process of broadcasting the radio channel from transmitter to receiver.

The quality of the received signal during the digital transmission practically does not depend on a distance from the transmitter, if the value of the radio signal exceeds

© The Author(s), under exclusive license to Springer Nature Switzerland AG 2022   1
M. Oziewicz, *Digital Radio DAB+*, https://doi.org/10.1007/978-3-030-66478-7_1

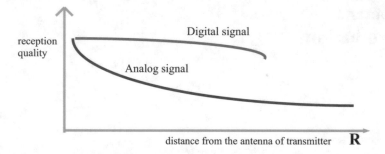

Fig. 1.1 The radio reception quality for analog and digital signals as a function of the distance from the transmitter

the minimum necessary to read the numerical samples of radio signal. Below this level, the reception quickly breaks down and turns into noise. Quality of analog signal is gradually degraded with increasing distance from the transmitter. These differences are illustrated in Fig. 1.1.

The digital signal can be protected against interference applying security and correction codes. In the case of analog signals, such opportunities do not exist. The prospects for better use of digital transmission channels by further optimizing signal compression methods, and the development of modulation techniques, are clear. Finally, digital broadcasting allows to transmit not only sound but also variable images and other data that can be presented in the digital form. Digital signal leads to extension of the classic role of the radio onto multimedia applications.

Digital radio receiver becomes the final part of the radio highway. Digital process allows to control the copyright of broadcasting programs by blocking the output jack of the receiver for programs covered by the license.

The innovative concepts used in the DAB+ digital radio enabled digital broadcasting by reducing the effects of interference during mobile reception, especially while driving a car. This is important because in developed countries the radio is mostly listened in the cars. Selected program can be listened to within the whole coverage area of the transmission network without the need for tuning the receiver when running from one transmitter to the next one.

Transmitter of digital system consumes less power than its analogue counterparts to cover with program the same area. Power consumption of a digital DAB or DAB+ transmitter is smaller than the consumption of an adequate FM transmitter.

New ideas of system DAB provided a foundation for concepts of digital terrestrial television and advanced mobile telephony.

1.2   Approaching Concept of Digital Radio

Digital radio DAB (Digital Audio Broadcasting) was originally planned as a traditional radio of higher quality. Inclusion of multimedia objects significantly broadened the DAB applications. This required an appropriate infrastructure on the

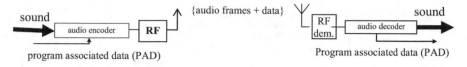

Fig. 1.2 Simple broadcasting channel for sound and associated data

transmission side and adequate functions of the terminal. Digital radio provides a higher sound quality in an open space. In the built-up or hilly areas, where a strong multipath propagation occurs, the use of simple broadcasting channel as in Fig. 1.2 is disturbed by overlapping signals. Therefore, to properly read the overlapping signals, there have been introduced on both the transmitter and receiver sides the specific mechanisms for securing reliable transmission for multipath, mobile reception [3]:

- New type of channel coder/modulator (OFDM)
- Detecting codes
- Correcting codes
- Time interleaving
- Frequency interleaving

These mechanisms are discussed in Chap. 2.

1.2.1 Frequency Blocks for DAB

In addition to resistance to the effects of multipath propagation, the DAB system is also planned for greater immunity to short-term electric disruptions. The interferences of such type will occupy the bandwidth of the FM channel. Maintaining the radio bandwidth for DAB as for FM system does not allow to reduce the impact of industrial noise during signal propagation, hence the idea to extend DAB channel to 1.5 MHz, which in case of disruption leads to only a partial loss of information recoverable by correcting coding system. For the adapted modulation method, the frequency block 1.5 MHz allows for throughput 1.5 Mbit/s. As the radio program after compression has several times smaller throughput, it leads to agreement for a simultaneous broadcast of several programs multiplexed into a single signal in one frequency block, like in Fig. 1.3.

1.2.2 Multimedia in the DAB+ System

The DAB+ radio as the digital system can be applied for transmission of the various types of data. The transmitted series of zeros and ones (binary system) may indicate the samples of sound, coordinates of the monitor points, the color parameters assigned to this point, codes of numbers or characters, etc. The DAB+ transmitter

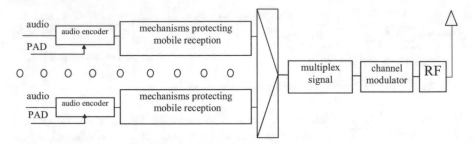

Fig. 1.3 Structure of the second-generation DAB transmitter

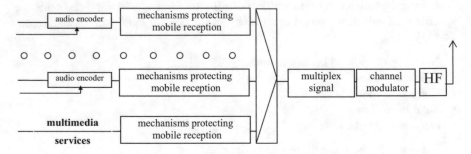

Fig. 1.4 Inclusion of the multimedia services in the DAB transmitter

can thus broadcast not only sound but also text, images, graphics, photos, and free video. To allow transmission of these audio independent media, there is necessary an additional subchannel aside from audio. Therefore, in the block diagram of the transmitters of new generation, there is assigned independent data subchannel, as shown in Fig. 1.4.

Several different monomedia (text, comment, picture, video) describing one event and thus linked by a common time and place of the intended exposure are combined in a single object called multimedia. DAB system is equipped with the mechanisms allowing for broadcasting such objects. Multimedia objects are transported in special files in accordance with the protocol MOT (Multimedia Object Transfer).

The data in the subchannel of digital information can be coded and compressed. Compression of still images is performed by the JPEG standard, motion pictures by MPEG, and text by, e.g., HTTP format. New versions of standards and new standards can be included.

1.2.3 Channels and Subchannels in the DAB System

Simultaneous transmission of several radio programs requires flexibility in the organization of the DAB+ signal transmission. The main transmission channel for audio and services, i.e., the Main Service Channel (MSC), with throughput of

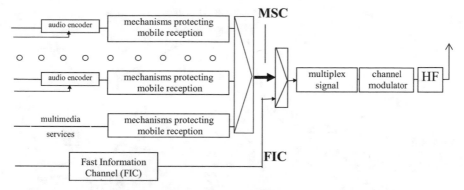

Fig. 1.5 The basic block diagram of the DAB transmitter. (After [1])

1.5 Mbit/s has logical frames divided into subchannels assigned to the individual operators of radio programs and data. This division may vary over time: transition from a broadcasting concert to reportage will be related with decreased throughput of the subchannel. The freed capacity of the subchannel can be assigned to another operator, i.e., for another subchannel. The flexible planning of the organization of the transmission requires a transfer of information to the receiver ahead of the planned changes in broadcast programs in order to activate an adequate decoder at the right moment. This requires the separation of information about the organization of the main channel in a special Fast Information Channel (FIC).

The basic role of the FIC depends on transporting information about the current and planned content of the subchannels to the receivers. In this way is organized the control system of DAB: information on throughputs and program contents of subchannels. This brings us to the basic block diagram of the DAB transmitter in Fig. 1.5.

In organization of subchannels, a data content of small capacity requires division of such services into packets – the basic building blocks of information with an assumed throughput of n × 8 kbit/s. To connect packets into original application in receiver, the data require addresses. The survey of programs and applications in receiver demands also specific labels, so these data also must be submitted in advance via the FIC channel to the receivers.

1.2.4 The Environment for Transmission and Reception of Multimedia Services

The transmission of multimedia services requires contact of the DAB server with different data sources from the various institutions in different locations. To this aim, the DAB system can take advantage of communication networks to gather data on its own server for multiplexing them in the form of a stream of audio associate data or

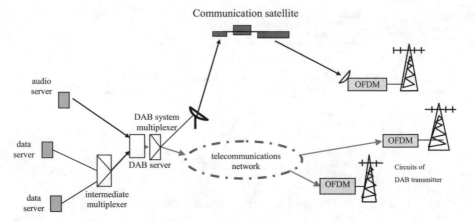

Fig. 1.6 Transmission of programs and multimedia services from sources of information to a network of transmitters

program independent data. This common multiplex signal is then transmitted via different communication networks (or satellite channels) to transmitters of DAB or DAB+ network(s). In transmitters the multiplex signal is encoded using redundant bits and modulated in the channel encoders generating a block of parallel frequencies called ensemble or OFDM signal, or in the case of single-frequency networks – the multiplex signal. Next, after passing high-frequency circuits, power amplifier, and output filter, OFDM signal is emitted by the antenna systems on the air.

Infrastructure for the selection, collection, and transmission of information to the DAB system server and next via the communication network to the network(s) of transmitters creates an environment for mobile multimedia broadcast, see Fig. 1.6.

References

1. F. Hermann, L. A. Erismann, M. Prosch, 'The evolution of DAB', EBU Technical Review, 2007
2. 'DAB+ in Europe: The Overview' https://www.dabplus.de, July 2020
3. 'Advanced digital techniques for UHF satellite sound broadcasting', Collected papers on concepts for sound broadcasting into the 21st century, EBU, August 1988

Chapter 2
Physical Layer of the DAB+ System

The role of the individual subsystems of transmitter is best illustrated through an example of a single local transmitter, where all elements are concentrated in one place.

2.1 The Block Diagram of a Local Transmitter

The block diagram of the local transmitter is shown in Fig. 2.1 [1].

The diagram highlights three main transport paths in the main service channel:

- The sound track with possible accompanying information in the subchannel PAD (Program Associated Data)
- The track of independent services in packet mode
- The track of independent services in stream mode

Every radio operator creates its audio or music program, alone or with accompanying data, organized in audio frames creating a subchannel of agreed throughput. Each subchannel is then independently protected by conditional access subsystem for paid information or data reserved only for authorized persons. Next follow energy dispersion subsystem, convolutional encoder, and finally time interleaver of encoder frames.

Security of transmission in each track is provided independently (not together for the joint signal), because in a receiver just one program can be selected, and only this portion of signal needs to be decoded applying slower processors, so the cheaper receiver.

Details on the different classes of consumer receivers are placed in Sect. 7.9.

Individual subchannels containing programs are then combined in a multiplexer (1) to form the Main Service Channel (MSC) of transmission. In parallel to audio subchannels with additional information related to the current programs, the channel MSC may include also the value-added services in a separate subchannel(s).

M. Oziewicz, *Digital Radio DAB+*, https://doi.org/10.1007/978-3-030-66478-7_2

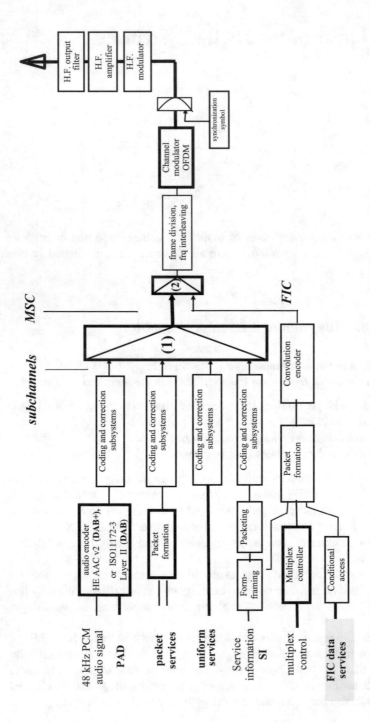

PAD – Programme Associated Data
FIC – Fast Information Channel
MSC – Main Service Channel

Fig. 2.1 The block system of a local DAB transmitter [1]

Information on the organization of subchannels within the common frequency block with the necessary parameters of time, location, or the terms of issue, and with Service Information (SI) describing programs, is sent to the receiver via an independent Fast Information Channel (FIC). Name of this channel is due to the mode of transmission providing information to the receiver ahead of actual programs and services to configure adequate decoders prior to delivery.

Fast Information Channel (FIC) is switched with the main transmission channel (MSC) via a further multiplexer (the second in Fig. 2.1).

At the output of the multiplexer, the signal is formatted in *logical frames*. Bits of logical frames are next projected onto the modulation scheme and, after frequency interleaving, are then used to modulate the channel encoder (modulator) of type Orthogonal Frequency Division Multiplexing (OFDM) creating *physical symbols*. The OFDM encoder is a part of the transmitter array that shapes the frequency and phase characteristics of the baseband signal.

Inclusion of the synchronization fields completes the construction of the *physical frames* of the DAB signal.

After passing high-frequency circuits, signal is emitted into the ether.

Below we discuss sequentially the action of the indicated subsystems.

2.2   Basic DAB Subsystems

Digital wireless systems are sensitive to noise, interference, and multipath. Therefore, regardless of the type of transmitted data, the modules play an essential role in each channel in protecting channel reception. Independently in each channel, before multiplexing all subchannels in one logical signal, as each channel can be selected and processed independently in the receiver.

2.2.1   Sound Track in Transmitter

What is the source of the compressed sound in the encoder it does not matter for its processing. However, the method of processing and data reduction results from the audio properties of listeners' ears (on the receiving side).

The sound track of a main service channel contains:

- Audio codec:

  - In the DAB, it is standard Music: ISO 11172 part three [2–4].
  - In the DAB+ system: standard HE AAC v2 (High-Efficiency Advanced Audio Coding) [5–7].
  - Both encoders include Program Associated Data (PAD) correlated with program

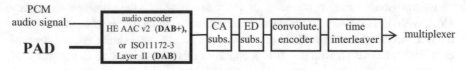

Fig. 2.2 Block diagram of the audio track in DAB or DAB+ transmitter [1]

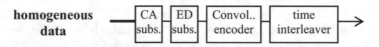

Fig. 2.3 Block diagram of the track of data stream services in DAB transmitter [1]

- Conditional access subsystem (CA)
- Energy dispersal subsystem (ED)
- Convolutional encoder (in receiver: Viterbi decoder)
- Time interleaver of the frames of encoder

The action of each of these modules is discussed in the following sections. Block diagram of the basic functional audio track in transmitter is presented in Fig. 2.2.

2.2.2   Track of Multimedia Services

Multimedia services can be transmitted either as a service accompanying the audio program –so-called Program Associated Data (PAD) services – or a service independent of the radio programs transmitted in an independent subchannel within the main service channel (MSC). Services of fixed bit rate and constant specific type can be sent in the stream mode. Services of different types or of variable capacity are sent in the packet mode.

Transmission of homogeneous data of fixed bit rate, such as video, can be accomplished through the DAB multiplexer after operations indicated in Fig. 2.3. Condition of such transmission is signal bit rate equal a multiple of 8 kbits per second. This requirement stems from the way of addressing the data in the DAB frames (see Sect. 6).

Services in packet mode are pre-connected (multiplexed) before entering the data subchannel, which also must have a throughput equal to a multiple of 8 kbit/s. Subsequent signal functional blocks in the packet mode proceed similarly (Fig. 2.4).

A conditional access system is here in two places to indicate possible coding at the level of services, its segments, or packages. The system of conditional access is discussed in Chap. 10.

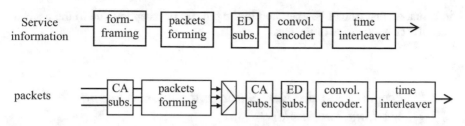

Fig. 2.4 Block diagram of the packet track in the DAB transmitter [1]

Synchro field	FIC field	MSC field

Fig. 2.5 Basic fields of the logical DAB frame

2.3 Basic Logical Frames of the DAB and DAB+ Systems

The logical signal in the DAB or DAB+ system, i.e., signal after multiplexes and before modulating the channel encoder, is formed in logical DAB frames of constant structure. Each DAB *logical frame* consists of three fields (see Fig. 2.5):

- Synchronization field (as part of Synchronization Channel)
- Fast Information Channel field (as part of Fast Information Channel, FIC)
- The main service channel field (as part of main transmission channel, MSC)

The output signal of transmitter is emitted in the form of *physical frames* shaped at the OFDM coder output. Operations of transforming the DAB *logical* frames onto the OFDM *physical* frames are described in Sect. 2.9.

2.4 Audio Encoder in the DAB System (MPEG II)

The purpose of an audio codec is compression of digital signal reducing outer bit rate, encoding the compressed samples, and formatting data in the standard outer audio frames.

The high-quality audio coding standard applied in the DAB system [2] is described by the norm ISO/IEC 11172–3 [3], which is a part of the MPEG-1 audio standard and norm ISO/IEC 13818–3 [4], known as MPEG-2.

The efficiency of an audio codec is indicated by the compression ratio. It is a ratio of the output bit rate to the audio bit rate at the input.

Because of the basic signification of the sound quality in radio, the method of compression in DAB is subject to special requirements.

Audio compression for broadcasting involves the removal of a redundant data from a digitally encoded audio signal in a real time. This redundant data includes information that cannot be exploited in the process of reception due to the psycho-acoustic characteristics of the ear.

2.4.1 Properties of the Ear: Static and Dynamic Sound Masking Thresholds

The human ear receives sounds of different frequencies with non-uniform sensitiv-ity. Testing ear response to the individual signal of a fixed frequency and increasing intensity (measured by relative air pressure oscillations around the ear), it appears that sound becomes audible only over a certain threshold. By changing gradually the frequency of the signal, one gets the values of the threshold of the hearing ear. To the average ear, the corresponding static masking threshold (as a threshold to mask the individual sounds) is shown in Fig. 2.6.

As tested, an average ear is most sensitive to sounds of frequencies from the range of 3–5 kHz. The corresponding intensity was taken as a reference for measuring the threshold for other frequencies. Masking threshold rises significantly for lower frequencies, alike as for frequencies approaching 20 kHz – an average limit of audible frequencies of the human ear.

In the case of simultaneous reception of sounds of close frequencies:

- Dominant harmonics form in their neighborhood *local masking zones* where the dominant harmonic components suppress reception of close frequencies.
- The masking zones of harmonic frequencies in close intervals do overlap.

The range of local masking zones depends on the intensity and frequency of the dominant harmonics. This is illustrated in Fig. 2.7a. Laboratory studies indicate that the properties of the local maxima in the spectral distribution of the incoming

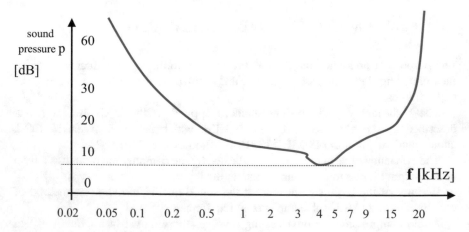

Fig. 2.6 The static masking threshold of an average human ear

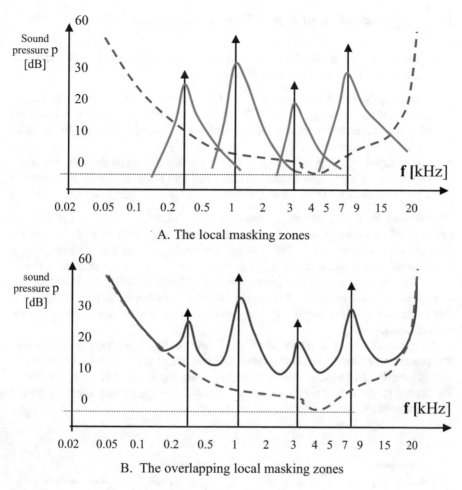

A. The local masking zones

B. The overlapping local masking zones

Fig. 2.7 The dynamic masking threshold

composite signal can have different meanings in masking process. The periodic components (tones) have a different shape of the masking curve than signals characterizing background noise sound.

During sound reception, the static diagram of Fig. 2.7a is subject to dynamic changes. Around the local maxima in the spectrum of the received signal are formed *dynamic masking* zones. Their total envelope defines the instantaneous dynamic masking threshold (see Fig. 2.7b). *Harmonics of a magnitude below this threshold are not perceived by the ear.* This property is used in the audio compression process.

2.4.2 Quantization of Audio Signal: Quantization Noise

The process of the audio signal from analog to digital – i.e., the Pulse Code Modulation PCM – consists of the following steps:

(a) Sampling the signal at a frequency of at least twice the maximum frequency of the signal spectrum f_M: $\Delta t = 1/(2f_M)$ (Nyquist sampling rate). Such sampling allows an accurate reconstruction of an analog signal, provided that the samples are perfectly accurate.
(b) Quantization, i.e., the assignment of each sample to one of the q levels. If the binary size of each level number is described by Q bits, the number of levels is $q = 2^Q$. This requires a bit rate $[1/(\Delta t)] \, Q = 2f_M Q$ bits per second.

The operation of quantization determines the samples of the signal with an accuracy of no higher than half the quantization interval (difference of the adjacent levels). This is illustrated in Fig. 2.8 for the signal both in time and frequency. Inaccuracies of quantization result in an effect of noise called *quantization noise*.

Obviously, the denser the quantization levels, the smaller the effect of the quantization noise. However, this requires a bigger number of bits per sample, and the higher the bitrate of the signal, the greater the frequency channel it requires for broadcasting.

It appears that the postulate of the high quality of signal reception is at odds with the postulate of limited signal throughput, if the signal is in the form of quantized digital samples. In the case of broadcasting systems, the limits of throughput result in the limited bandwidth of frequency channels. It follows that reduction of transmitted data is necessary.

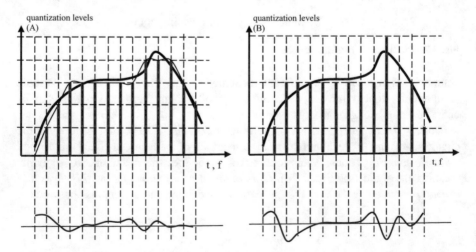

Fig. 2.8 The analog signal, digital signal samples, and (lower) the quantization noise depending on the density of quantization levels

In the DAB encoder, the quantization noise is masked using the psychoacoustic properties of the human ear.

2.4.3 Masking Quantization Noise: Sub-band Encoding

Maintaining high-quality quantized audio requires a reduction of quantization noise below the threshold of a hearing ear. The characteristics of the masking threshold depend on the frequency (Fig. 2.7), so the signal bandwidth is divided into many sub-bands, and in each of them, its own quantization noise level is analyzed. It is determined by minimum threshold of hearing in sub-band (Fig. 2.9). This allows to optimize the quantization process by reducing the number of quantization levels permitted in individual sub-bands by individual levels of quantization noise.

In the MPEG encoder, the audio bandwidth is divided into 32 sub-bands of 625 Hz. Digital filtration of input samples in batches of 1152 samples (input PCM frames) is performed using a polyphase filter algorithm. In this way we obtain in each of the 32 sub-bands 36 samples in 16-bit binary code (2^{16} quantization levels). The compression process depends on reduction of the size of this code independently in each of the sub-bands in such a way that the quantization noise in sub-band does not exceed its dynamic masking level.

Digital filtering of input signal divided into sub-bands leaves unaltered the total number of samples. The possibility of reducing the number of levels in sub-bands (i.e., bits per sample) leads to compression of the output signal.

Action of the encoder is illustrated in Fig. 2.10. Analog input signal is sampled in analog-digital converter with a frequency of 48 kHz and an accuracy of 16 bits per

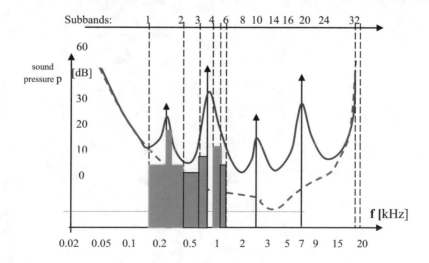

Fig. 2.9 The acceptable levels of quantization noise in the sub-bands

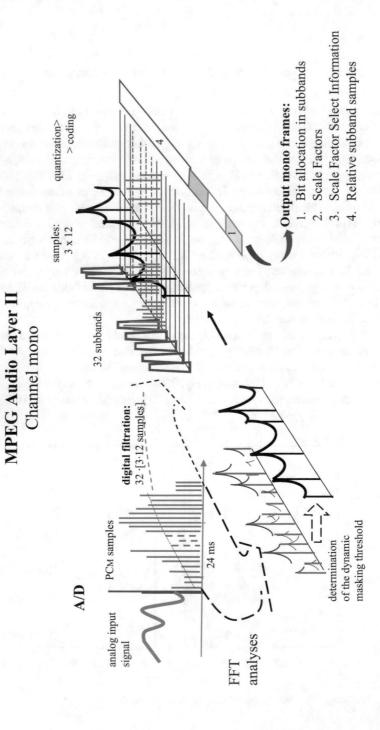

Fig. 2.10 The main steps of processing in the DAB encoder

sample. This means that the amplitude quantization relates the samples into one of 2^{16} levels. The bitrate of this signal is 768 kbit/s.

The samples of the signal from AD converter are divided into groups sequentially undergoing filtration forming digital audio signal frame. Each group goes through parallel processing in procedures:

• Determination of the current dynamic quantization threshold

The group of samples (batch) is transformed by fast Fourier transform (FFT) giving the spectrum in the analyzed period 24 ms. This allows to define masking zones of the dominant harmonics in sub-bands, which serve for constructing dynamic masking threshold in accordance with the rules stated in [2]

• The digital filtration separating the signal into 32 sub-bands

Samples are converted in a bank of 32 filters to 32×36 samples, 36 samples in each sub-band. This is the basis to reducing the number of quantization levels sufficient to mask the noise in sub-bands.

• The optimization of the number of quantization levels

The optimized dynamic masking threshold in acoustic frequency range is used to calculate the minimum masking thresholds in the sub-bands. This allows the optimization of the number of quantization levels and sampling bit allocation in sub-bands.

2.4.4   Organization of the Outer Frame of the Encoder

The DAB encoder offers the following modes for audio compression [2]:

• Mono in single channel
• Dual channel (two mono channels)
• Stereo mode
• Joint sterco
• Multi-channel

Each of these cases is described in the following. As a result of signal processing in encoder, a sequence of output frames is formatted. Every frame contains compressed data of input samples covering 24 milliseconds. Since the coding takes place in real time (on-line), so also each outer frame should last 24 milliseconds. Capacity of the frame depends on the assumed signal compression.

Control of output bitrate is set by the operator of the service. Bit allocation for a single frame follows this assumed bitrate. Next, the algorithm of the encoder determines the allocation of bits for coding parameters and optimizes the bit allocation for quantization the samples in each sub-band.

2.4.4.1 Determination of the Basic Parameters of the Frame

The basic steps in transcoding input mono signal in PCM code into the outer signal formatted according to the DAB specification [2] are described in Fig. 2.10 and the algorithm in Fig. 2.11.

- **The measurement of scale factors and their choice in each sub-band**

The sound dynamics from 0 to 120 dB every 2 dB is determined in each sub-band by the maximum samples called the scale factors. The scale factors are subject to special protection, what is facilitated when they are isolated in a separate field of the outer frame. The scale factors are measured as the maximum absolute values in three successive groups of 12 samples in each sub-band. Each factor is encoded by a 6-bit word and assigned to the next higher quantized level. Based on the relation between the three scale factors of each successive groups in sub-band, not all factors are transmitted. This is recorded in the parameter Scale Factor Select Information, placed in the field next after scale factors.

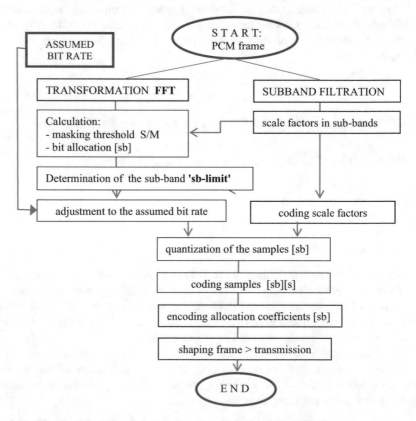

Fig. 2.11 The algorithm of coding the mono channel for levels I and II [2]

- **Calculation of the relation of dynamic masking threshold level (M) to quantization noise (n_q) in the sub-bands**

  Procedure of masking the quantization noise considers the parameters:

  1. The current masking threshold M (M value corresponds to the minimum dynamic masking threshold in the sub-band)
  2. The quantization noise n_q in each of the sub-bands

  Relation M/n_q is calculated from the difference (in dB) $S/n_q - S/M$, where S is the level of the current signal in sub-band. Depending on the number of quantization levels and the quantization model, the relation S/n_q can be calculated a priori. Maximum signal S is calculated for each sub-band at the output of the digital filter. System specification [2] gives values n_q in the table as function of the number of quantization levels.

  The quotient S/M is the dynamic value changing in course of time with each frame and each sub-band and is successively computed. Masking threshold level is calculated from a spectral distribution of 1024 input samples. After separation of the tone (quasi-periodic) and noise-like components, which have different shapes of individual masking thresholds, the value of masking curve at analyzed point is derived from adding the value of the static threshold and the sum of individual masking thresholds of close harmonics in this point. Effective masking of quantization noise requires that the $M/n_q > 1$.

- **Bit allocation for the individual sub-bands for encoding the samples**

  The capacity of the coder output frame, so the number of available bits, depends on the predetermined output bit rate. Allocation of bits to individual sub-bands determines the quality of the outcoming sound, i.e., the quality of the encoder. The optimum choice is calculated in an iterative way. As shown in Fig. 2.8, the more the quantization levels, the smaller the quantization noise. Hence the strategy is to conduct the allocation of quantization levels (code bits) in the sub-bands in such a way that relation of the masking threshold to the quantization noise (M/n_q) in sub-bands was together as large as possible, but does not result in exceeding the allocation of available bits in the frame. After the adoption of the initial number of quantization levels per sub-band (e.g., [1]), the iteration steps include the following actions:

  - For each sub-band calculate: M/n_q [dB] $= S/n_q - S/M$.

    where the first term depends on the assumed number of quantization levels, while the second shows the dynamic masking threshold obtained from the FFT analysis.

  - In the sub-band with the lowest value of M/n_q (the highest level of the quantization noise in relation to the masking threshold level), the number of quantization levels is increased reducing the level of quantization noise. Resulting increased demand for bits in sub-band is next added to the initial value.
  - If the number of free bits in the frame is not exceeded, renew the procedure.

Data needed in the iteration steps are calculated using the tabulated coefficients in [2] arising from heuristic data of acoustics. If the number of bits for encoding sub-band samples exceeds the assumed limit, allocation of bits in this sub-band is assigned a value of zero. In this way, the samples below masking threshold are omitted.

2.4.4.2 Signal Mono

Following the procedure described above, the samples from the highest sub-bands are not always transmitted. For each frame, the allocation of bits for sub-bands above the determined limit called *sb_limit* (< 32) is equal to zero. The effective range of the sub-band index *sb* is therefore in the range from 0 to "sb_limit." The outer frame of encoder for mono signal contains the following basic fields:

- *Bit allocation*

This field contains the allocation of bits for encoding the samples in each sub-band.

- *Scale factors*

Samples with a maximum module in each sub-band (for scaling the remaining samples).

- *Scale factor selection*

The index of scale factors across individual sub-bands.

- *Codes of sub-band samples*

Here are placed the codes of samples related to the scale factors in sub-bands.

Code "the scale factor selection" in the sub-band *sb* determines the amount of scale factors contained in this sub-band. A detailed description of the system is contained in system specification [2]. For the sub-band index "sb" in the range $0 \leq \text{sb} < \text{sb\_limit}$, the fields of frame are shown in Fig. 2.12.

2.4.4.3 Signal Stereo or Two-Channel Mode

This mode is used to compress audio with two independent channels (e.g., translations in various languages, the sound accompanying the image + comment) or sound stereo. Frame mode for dual or stereo channel is shown in Fig. 2.13. For

Bit allocation [sb]	Scale factors [sb]	Scale factor selection [sb]	Codes of sub band audio samples [sb]

Fig. 2.12 Organization of the mono signal frame

Bit allocation [k][sb]	Scale factors [k][sb]	Scale factor selection [k][sb]	Codes of subband audio samples [k][sb]

Fig. 2.13 Frame fields of the two-channel or stereo signal

indications as above, and the channel number of the index 'k' (k = 0, 1) – frame fields give performance of the first, then the second channel.

2.4.4.4   Compact Stereo Mode

Particularly strong compression of stereo signal can be achieved at the cost of slightly greater complexity of the encoder signal processing. Comparing the samples of signals from both channels in each sub-band eliminates the double transfer of codes of correlated signals. This applies to the higher sub-bands, from the fixed sub-band named "isl" (intensity stereo limit) in each frame. The number of this sub-band is given in the frame header. The data below this value on the two channels are independent; above it is the sum of the signals sent from both channels. Only the scale factors for both channels are given independently. Each of the fields of normal stereo mode is therefore divided into two parts. The first, describing individually the two channels, includes sub-bands numbered *0 – isl*, and the second treats the two channels jointly, from *isl* up to *sb_limit*. Schematically it is shown in a sketch of Fig. 2.14.

The range of other parameters is the same as in normal stereo mode (Fig. 2.15).

The values of codes in each field depend on the signal waveform and frequency of sampling and determine the allocation of bits per frame according to the bitrate index.

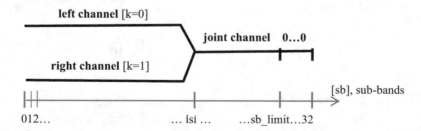

Fig. 2.14 Schematic organization of the sub-bands of compact stereo mode

bit alloc. [k][sb]	bit alloc. [sb]	scale factors select. [k][sb]	scale fac-tors select. [sb]	scale fac-tors [k][sb]	scale fac-tors[sb]	Codes of audio samples[k][sb]	Codes of audio samples [sb]

Fig. 2.15 Fields of the compact audio frame [2]

2.4.4.5 MUSICAM-Surround

MUSICAM-Surround (sound surround) is a versatile multichannel encoding system
compatible with the two-channel system described in ISO 11172-3 [3, 4]. The
auxiliary data field in the frame of MPEG-Audio has been used in this system for
the multichannel enlargement. Three additional channels can be used to transfer the
translation of dialogue program in additional languages and additional sound effects
or to extend additional stereo transmission channels (total of five). In the latter case,
the basic stereo channels – left and right – are supplemented by an additional center
C channel and two surround channels, left and right, Ls and Rs. Such system is
referred to as "3/2 stereo" (3 front/2 surround channels) (Fig. 2.16).

Compatibility of the sound surround with the standard MUSICAM layer II in the
compact stereo mode is marked in Fig. 2.17. This ensures "the compact encoding
mode" in which linear combinations of the five channel signals $L/C/R/L_s/R_s$ produce
a basic stereo signal L_o/R_o and additional subchannels T3/T4/T5 according to
equations:

$$L_o = L + xC + yL_s$$
$$R_o = R + xC + yR_s$$
$$T_3 = C$$
$$T_4 = L_S$$
$$T_5 = R_S$$

where x and y are the ratio of the central and surrounding signals to the basic one.

Relation of the transmitted channels to the original ones is shown in Fig. 2.17.
The signal bitrate required for multichannel system is not proportional to the number

Fig. 2.16 Location of
speakers in the 3/2 stereo
system

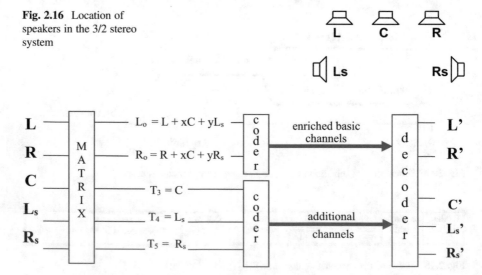

Fig. 2.17 The idea of the MUSICAM-Surround system

of channels. The reduction rate is based on psychoacoustic properties of reception of extended stereo channel:

- Certain parts of the stereo signal not correlated with the localization can be transferred by any speaker.
- Some stereo signals containing coherent elements can be transmitted over a single channel.
- The effect of a cross-channels audio masking allows samples below the masking threshold in the other channels to be eliminated.

The total bitrate required for a MUSICAM-Surround system is 384 kb/s. The division into basic and additional channels allows several options of bitrate division (256/128, 224/160, 192/192, 160/224). Selection may depend on the level of the desired multichannel transmission, or the required number of channels.

MUSICAM-Surround frame corresponds to the standard ISO 11172-3 [3]. In addition to the header, CRC, and data fields, the complementary field is implemented for transmission of data associated with channels T3/T4/T5 (Fig. 2.18). Two bits reserved in the frame header would include information on the size of this field. The decoder compliant with ISO 11172-3 for the two-speaker configuration should ignore this piece of information.

Recommended bit rates of MUSICAM encoder, depending on the operating mode, are given in [2] (Table 2.1).

2.4.4.6 Decoding Algorithm

The basic steps of the algorithm for decoding a mono channel are shown in the Fig. 2.19 [2].

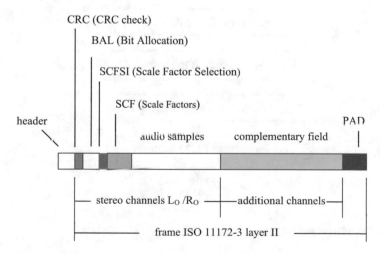

Fig. 2.18 The frame of encoder for MUSIC-Surround audio system [2]

Table 2.1 Recommended bit rates of compressed audio signal

Bit rate [kbit/s]	Modes of encoder processing			
	channel channel	channel joint	mono double	stereo stereo
32	x			
48	x			
56	x			
64	x	x	x	x
80	x			
96	x	x	x	x
112	x	x	x	x
128	x	x	x	x
160	x	x	x	x
192	x	x	x	x
224		x	x	x
256		x	x	x
320		x	x	x
384		x	x	x

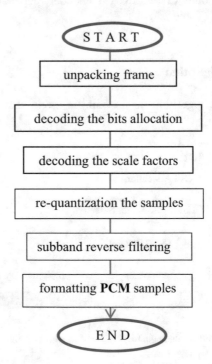

Fig. 2.19 Decoding algorithm for mono channel frame, level I

After decoding the header of the frame, the next fields are unpacked and decoded.

In the case of joint stereo mode, the intensity stereo limit is considered. Code control runs the CRC checking procedure. Detection of error causes substitution of the frame by the previous frame to avoid distortion of sound.

Knowledge of the bit rate and sampling frequency allows to decode the bit allocation for the individual sub-bands using tables of standard. Thus, the amount of quantization levels is determined, and next is settled the number of code bits of the grains or individual samples.

For sub-bands, where the allocation of bits is non-zero, samples are decoded in sequence:

the choice of scale factors → scale factors → the samples

Reverse filtration reconstructs the output samples of the PCM decoder.

2.4.5   Program Associated Data (PAD): F-PAD and X-PAD

Multimedia radio DAB or DAB+ opens the possibility of transmission multimedia correlated with services. These services are transmitted in data subchannel with acronym PAD (Program Associated Data) accompanying the audio program. Subchannel PAD is created by special fields in the audio frames.

2.4.5.1   The Dynamics of Filling Frames

Audio frames have a fixed volume determined by the selected encoder output bitrate. However, filling the frames with audio data resulting from the applied algorithm varies from frame to frame depending on the current signal parameters. Therefore, in the audio frames, there remain fields not effectively fulfilled. A change in the size of the free space in the consecutive frames is illustrated in Fig. 2.20.

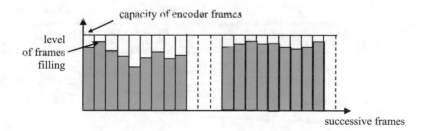

Fig. 2.20 The dynamics of filling the MUSICAM audio frames

Practically in each MUSICAM frame, a field carrying additional information can be organized without reducing sound quality. These fields form a Program Associated Data channel, PAD, accompanying service.

2.4.5.2   Organization of Services Accompanying the Audio Program (PAD)

A field at the end of each frame not filled with bits of audio code creates a data field PAD accompanying the program. A total of PAD fields in all audio frames create subchannel PAD. It can be applied to provide services synchronized with programs that do not require addressing, e.g., text accompanying the song, the picture of performer, etc., or information independent of the program.

In the case of transmission of data associated with the program, the chunks of data in the PAD frame "n" are correlated with the sound (audio) in the next frame "n + 1", as shown in Fig. 2.21:

The PAD field consists of two parts, F-PAD and X-PAD:

I. A *constant field F-PAD (Fixed PAD)* at the end of each audio frame, of fixed 2-byte length. In mode I, the throughput of the channel F-PAD is equal to 0.667 kbit/s (2B per CIF) × (4CIF per frame) × (10.42 frames per sec).

Depending on type of the F-PAD field, it can comprise (Fig. 2.22):

> The length indicator of the next variant field X-PAD with options:

 – No X–PAD field
 – The fixed length of the X-PAD equal to 4 bytes in each audio frame
 – A variable length of the X-PAD from frame to frame

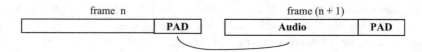

Fig. 2.21   The data associated with the audio frame (n + 1) are found in the preceding frame n

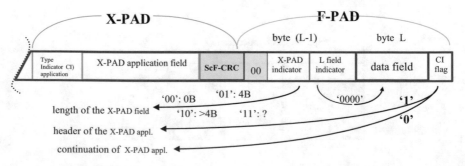

Fig. 2.22   F-PAD (Fixed-PAD) field configuration for F-PAD type "00" [2]

> The contents of the last byte L that could fit:

- Support of own technical information of the system called in-house information
- Control of signal dynamic in the next frame in the range 0–15.75 dB in steps of 0.25 dB
- Optional code indicator: music or voice

> The flag CI (Content Indicator) specifying the beginning ("1") or continuation ("0") of applications in the optional X–PAD.

Resetting the F-PAD field means the lack of information in the PAD.

Each part of the field F-PAD and the location of parameters of the X–PAD field are indicated in Fig. 2.22 [1].

II. An *extended part of PAD (X-PAD)* is contained between the last sample of the audio and the scale factors error check (SCF-CRC) preceding the F–PAD field.

The length of the X-PAD is established by the service operator. The length indicator field of X-PAD (value 0 means the lack of the field) is placed in the F-PAD. The X-PAD field is divided into an application field and – only if the header is transmitted – an indicator of the type of application CI (Content Indicator), which specifies the type of encoder required in the receiver for decoding X-PAD application. For applications that are transported in several successive X -PAD fields, the information about the beginning of the application or its continuation is determined by the flag CI.

Applications of large capacity can be described as data groups or multimedia objects MOT (Multimedia Object Transfer) described in Chap. 7. In one field of X-, PAD up to four different applications can be transported simultaneously. In this case, the content type indicator is required for each application.

In the X-PAD fields of variable length CI indicator also gives the length of the X-PAD, and the distribution of the X-PAD among several applications, if such option is taking place.

Among permitted number of 31 different types of applications and transport indicators given in the index CI, the currently defined types are listed in Table 2.2. Depending on the segmentation method of the application (segments, data groups, media objects MOT), the type of application is marked in the channel X-PAD.

For example, the dynamic label can consist of up to 8 segments transmitted in the consecutive X-PAD fields, each containing up to 16 characters.

Transmission services in subchannel PAD can be performed in one of three modes:

- The Stream Mode, where one subchannel contains one application.
- The Packet Mode which allows up to 1023 different types of packets for subchannel. The length of packet ranges from 24 to 96 bytes. Packets of one program are identified by an established address. The maximum number of simultaneous applications in this mode is equal to 64.

Table 2.2 The applications in channel X-PAD covered by the DAB specification [1]

Application type	Description
0	End marker of application
1	Data group length indicator
2	Dynamic label segment, start of X-PAD data group
3	Dynamic label segment, continuation of X-PAD data group
4–11	User defined
12	MOT, start of X-PAD data group packets
13	MOT, continuation of X-PAD data group, see [6]
14	MOT, start of CA messages
15	MOT, continuation of CA messages
16–30	User defined
31	Actually not used

2.5 Audio Encoder in the DAB+ System

The main difference between the systems DAB and DAB+ is the introduction in the latter the audio encoder with a much higher level of performance and flexibility with the possibility of transmission in frequency channels of different widths. Encoder HE AAC v2 (High-Efficiency Advanced Audio Coding) used in DAB+ [5–7] is made up of selected components of MPEG-4 audio, in which a combination of different mechanisms of audio coding can be chosen. For the DAB+ applications, three elements were selected from a set of MPEG-4 audio:

A Advanced Audio Coding (AAC)
B Spectral Band Replication (SBR)
C Parametric Stereo (PS)

Ad A. The MPEG AAC encoder works for throughput of about 175 kbps in the case of an audio signal with no accompanying information. Just as the MUSICAM encoder, AAC applies signal representation in the frequency domain. The basic steps of encoding are the same as the encoding mechanisms of each psychoacoustic encoder:

• The spectral analysis of the signal
• The effect of masking. Determination of masking thresholds
• Filtration and signal encoding in sub-bands
• The bit allocation and quantization in sub-bands

Efficiency of AAC encoder is higher than the performance of the MUSICAM encoder. In addition to various forms of prediction, it stems mainly from the application of transforming blocks of varying lengths, depending on the location of the analyzed acoustic frequency range. Transformed blocks in the most sensitive range of 3–5 kHz count of 256 samples, where in the lower and higher ranges, it reaches 2048 samples.

The encoder MPEG AAC adopted a sampling rate of 32 kHz or 48 kHz. In the basic Access Unit (AU) frame, the number of samples is 960 per channel.

Ad B. Spectral Band Replication (SBR) concept depends on reconstructing the upper part of the spectrum of the audio signal with the help of a suitably parameterized replication of the bottom part of the spectrum. This saves about half the capacity of the code of signal encoded correctly. The SBR is based on the information from the decoded signal output from the MPEG AAC coder band and a small number of basic control data to ensure optimum high-frequency reconstruction. Limiting the analysis to the lower-frequency band can be obtained by operating with the sampling rate reduced to half, i.e., 16 or 24 kHz, for upper part of the spectrum. For parameterization of this part, the original frequency is held. At the upper part, the spectral envelope is estimated using the set of QMF (Quadrature Mirror Filters), for establishing SBR parameters. After quantization – considering psychoacoustic model – SBR parameters are encoded and included in the code stream of the lower part of the spectrum.

Processing in decoder starts with decoding the lower part of the spectrum, followed by moving the replica of this part to the place of the high band and next matched to the original size using the SBR parameters. The basic steps of action on SBR encoder side and decoder side are shown in Fig. 2.23.

Ad C. The Parametric Stereo (PS) method depends on the reconstruction of stereo channels based on knowledge of the basic signal flow and the parameters distinguishing individual stereo channels. PS coder operates at the SBR block as additional information (side information). Parameter code of audio signals uses a

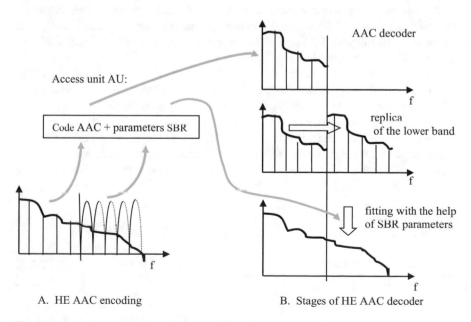

A. HE AAC encoding   B. Stages of HE AAC decoder

Fig. 2.23 Idea of the HE AAC mono channel encoder [7]

technique called HILN (Harmonic and Individual Lines plus Noise) for throughput around 4 kbit/s or higher.

In the HILN encoder, the sinusoidal waveforms, harmonic tones, and noise are adequately parameterized and are used for representing different versions of the source signal. Parameters are quantized considering the psycho-perceptual delivery model. In place of independent code for each channel, only mixed signal source is transmitted with a small number of parameters relevant to the reception quality of the diverse individual stereo channels. Such procedure significantly increases the efficiency of the encoder.

Knowledge of HE AAC v2 code allows to reproduce either the full code of AAC channel or the HE AAC code, which allows for the flexible use in the receiving devices with different profile of equipment.

Depending on the bandwidth of the channel, only the AAC encoder, AAC encoder with SBR replica of the spectrum, or the sum of all mechanisms can be applied. In the latter case, the performance of the encoder is equal to about one-third of the efficiency of the MUSICAM encoder.

The encoder is used to compress the audio signal:

- Mono
- Stereo
- Multichannel sound surround system *MPEG Surround*

In the multichannel system, *MPEG Surround* individual input channels transmit mono or stereo audio signals and associated special parameters. The output stream is constructed from mixing the input channels and identifying parameters allowing to distinguish the differences between each of the individual channels. Depending on the software of the receiver and its ability to run additional optional tools, it is possible to distinguish between different profiles and levels defining different configurations of the encoder (Fig. 2.24).

The frame transferring encoded data and parameters is called the Access Unit (AU). The length of the Access Unit (AU) in the system DAB+ depends on the type of encoder and the sampling frequency, in accordance with Table 2.3.

The construction of the output frame of encoder is unified to simplify the transmission and decoding in the system DAB+. The carriers of the audio are super frames with length of five DAB logical frames. Super frames are a common base for different encoder profiles and different sampling frequencies, requiring different lengths (Fig. 2.25).

Fig. 2.24 Concept of multichannel audio transmission

Table 2.3 Variant parameters of MPEG 4

Type of coder	Sampling frequency [kHz]	Number of samples/channel	Access Unit (AU) [ms]
AAC	32 or 48	960	30 or 20
AAC core + SBR	16 or 24	960	60 or 40

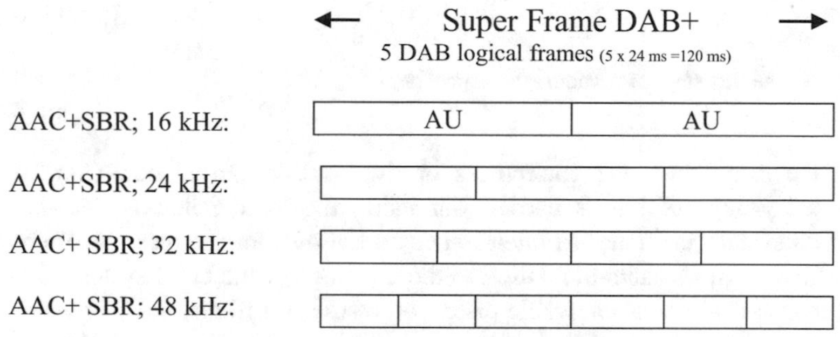

Fig. 2.25 Construction of the super frames for different Access Units (AU) [5]

Parameters of super frame header contain information about the type of encoder and the frequency sampling rate:

- Sampling rate of converter ΛC (32/48 kHz)
- The presence of SBR encoder (yes/no)
- AAC encoder type: mono/stereo
- The presence of PS encoder (yes/no)
- The basic configuration of the MPEG Surround and parameters identifying configuration
- The addresses of Access Units (AU) in bytes

The number of received channels in MPEG Surround depends on the consumer and is limited by the technical equipment of the receiver. Required information is included in the parameter SpatialSpecificConfig. The frame fields of the HE AAC encoder are protected by Reed-Solomon code common for the whole super frame.

At the receiver, decoding begins with the identification of a super frame header. For this purpose, at least five successive DAB frames are loaded into a buffer for viewing the synchronization fields parallel with moving consecutive frames and replacing the buffer with new frames. When reaching the super frame synchronization field, the decoding cycle begins:

- Super frame synchronization header
- Downloading audio encoder parameters
- Reading addresses of Access Units (AU(n)) in super frame
- Reading parameters from the extension frame (Fig. 2.26)

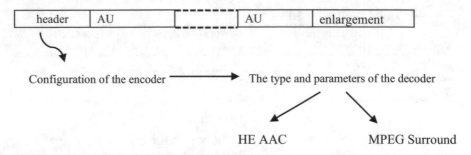

Fig. 2.26 The first steps of decoding the super frame

At the same time, the correctness of the code is carried out. Correction of corrupted groups of bits is carried out using the Reed-Solomon decoder. This increases system resistance to interference and allows for the increase of coverage area of the DAB+ system in relation to the coverage of the DAB system.

In the case of a non-removable distortion of the AU frame, the corrupted one is replaced by an average of the correct data from the boundary AU. Averaging is carried out at the level of the scaling factors.

Each AU frame may include Program Associated Data (PAD) just like the MUSICAM system.

2.6   Mechanisms Securing Quality of Transmission

The DAB system is designed for the transmission of digital data in broadcasting channels for fixed and mobile reception. Due to possible interference of the channel by:

– Multipath propagation
– Doppler shifts
– Industrial noise or static electricity

reduction of error rate below an acceptable level requires the use of the security mechanisms. In the system DAB, there are applied mechanisms:

• The CRC codes used for *error detection* of frames, packets, and data group. These units cannot be too long to avoid too much data portion in the fault detection.
• A convolutional encoder for *error correction* used in the transmitter and a Viterbi decoder at the receiver. The extension of the encoder efficiency is achieved by introducing a mechanism of puncturing.
• In the case of *serial errors*, caused by interference with a duration longer than 300 microseconds, Viterbi decoder is not working effectively. To overcome this

difficulty, a frame time interleaving decoupling the bits of one frame between the 16 successive frames was used.

• With channel modulation in DAB and DAB+, the Orthogonal Frequency Division Multiplexing (OFDM) system has been applied. This modulation system developed especially for digital radio plays a basic role in overcoming the problems of multipath propagation. This system is discussed in Sect. 2.9.

2.6.1   The CRC Codes of the DAB Frames

A code CRC (Cyclic Redundancy Code) is a classic code allowing detection of errors arising during transmission [8].

Every finite sequence of bits can be treated as a system of polynomial coefficients. So instead of the bit transmission, it can be equivalently regarded as sending the corresponding polynomial.

The essence of the CRC code depends on comparison of the remainder of division of the transmitted polynomial by a priori known on the transmitter and receiver sides the fixed so-called generating polynomial. If the transmitted string will not be distorted and thus misrepresented, the remainder of the division on the sending and receiving sides should be equal. Otherwise, we have transmission with errors.

To implement this scheme, it is necessary – besides the transmitted data – to send also the remainder of division in the transmitter to be able to compare it with the remainder of the division in the receiver. Division is performed using the shift registers.

Problems with the CRC generating polynomial depend on a choice that leads to:

• Unambiguous result of division (it cannot be a priori excluded that different polynomials divided by the same generating polynomial give the same rest).
• The detection also of successive double and triple errors
• Possibly a short remainder (the fast transmission of remainder)

Optimization of the CRC codes stems from the theory of algebraic polynomials.

In telecommunications, the selection of generating polynomials has been standardized.

In the DAB+ system, the following polynomials are used:

• $G_1(x) = x^{16} + x^{12} + x^5 + 1$ (ITU-T Recommendation X.25).
• $G_2(x) = x^{16} + x^{15} + x^2 + 1$ ('CRC-16').
• $G_3(x) = x^8 + x^4 + x^3 + x^2 + 1$ ('CRC-8').

The G1 polynomial is used to calculate the remainder of division from:

– The data fields in the frame of Fast Information Blocks (FIB) in Fast Information Channel
– The header and the data fields of Main Service Channel (MSC)

– The headers (group and session) and a data group in the transport layer of main service channel

The G2 polynomial is used to detect errors of transmission in the fields of "Scale Factor Select Information" in audio frames.

The G3 polynomial is used to detect errors of transmission in Program Associated Data (PAD) fields in audio frames.

The response to the signal of errors of transmission depends on the type of the received information and is realized by the system software.

2.6.2   Convolutional Encoder with Puncturing

The aim of the convolutional encoder is correction of transmission errors in real time.

The encoder algorithm corresponds to the hardware version implemented as a shift register composed of L flip-flops connected in cascade with serial input. Codeword **x** containing n bits is produced at the output of n linear combinations of selected outputs of flip-flops, e.g., in Fig. 2.27 for n = 4.

During the process of coding, successive bits ai of information implement assigned code words **xi** containing a larger number of bits than the input bit sequence. The ratio of the number of entering bits to the outgoing codeword bits is defined as the code efficiency. During transmission, some bits of code can happen corrupted. In this case, codeword redundancy increases the probability of choosing in this place the correct codeword by identifying the most likely codeword from a finite set (library) of valid code words known in receiver. To this aim it is necessary to estimate the channel and so probability of distortion introduced into the codeword during transmission. In the receiver the operation of decoding is repeated sequentially for each received codeword. Indication of the most likely transmitted code allows to recreate an adequate series of information bits.

The algorithm applied in the DAB system for decoding the convolutional code was developed by Viterbi [9]. It allows to determine the most likely succession of permissible consecutive bits of redundancy code from the received string possibly containing the corrupted bits.

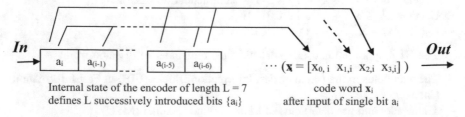

Internal state of the encoder of length L = 7   code word x_i
defines L successively introduced bits $\{a_i\}$ after input of single bit a_i

Fig. 2.27 Implementation of the hardware model of convolutional encoder

In order to optimize the bitrate of the coded signal, not all the content of the input signal is protected with the same efficiency: less protected content is encoded with codewords with fewer bits. Reducing the number of bits in indicated codewords is obtained by process of puncturing. The puncturing depends on a selection of a priori established bits in the outgoing codewords and ignoring the others. Another term for this process, used in parallel, is perforation.

Punctured convolutional code was selected among other codes because of the DAB requirements:

– Simple decoder easily implemented as software realization (soft decoding)
– The ability to adaptive changing efficiency of the code in real time

The last property provides different levels of protection for the selected fields in audio frames and different packets depending on their importance for the transmitted content.

The software implementation of the algorithm depends on designating codewords based on the actual internal state of the encoder. In this way the so-called mother code is generated. The mother code in the DAB system is defined by the following generating polynomials [1]:

$$a_i + a_{i-1} + a_{i-2} + a_{i-3} + \qquad +a_{i-6} = x_{0,i}$$
$$a_i + a_{i-1} + a_{i-2} + a_{i-3} + a_{i-4} + \qquad +a_{i-6} = x_{1,i}$$
$$a_i + a_{i-1} + a_{i-2} + a_{i-3} + \qquad +a_{i-5} + a_{i-6} = x_{2,i}$$
$$a_i + a_{i-1} + a_{i-2} + a_{i-3} + a_{i-4} + \qquad +a_{i-6} = x_{3,i}$$

The output vector x_i is defined by linear combinations of bits

$$a_{i-6}, a_{i-5}, \ldots a_{i-1}, a_{ir}$$

which determine the internal state of encoder. Each successive input bit a_i generates its own output vector x_i, depending also on the earlier inputs. The initial state of the encoder corresponds to the zero values of the internal bits. The final state of encoder, after insertion of all the input data bits and next successively filling free places with zeros, creates six vectors, $x_{i+1},..., x_{i+6}$, which correspond to the 24 bits of the so-called tail.

In the DAB system, the outgoing codewords are organized into groups and subgroups adjusting the code structure into the frames. The aim of the grouping operation is to prepare the outgoing sequence to the puncturing process. The perforation determines bits of code intended for transmission. It is a simple way to realize the adaptive efficiency of encoder. For this purpose, vectors x_i are serialized with increasing index "i" creating a string of bits $\{u_i\}$:

$$x_0, x_1, x_{00}, x_{10}, x_{20}, x_{30}, x_{01}, x_{11}, x_{21}, x_{31}, \ldots x_{nm}, \ldots =$$
$$= u_0, u_1, u_2, u_3, u_4, u_5, u_6, u_7, ..., u_{4m+n}, ...$$

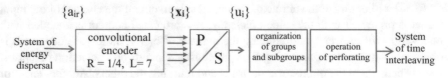

Fig. 2.28 Functions of convolutional encoder with perforation

The sequence $\{u_p\}$ is divided into groups of 128 bits (32 consecutive vectors x_i). Each group is decomposed into four subgroups of 32 bits (8 consecutive vectors x_i).

Perforation is tailored to the organization of Fast Information and Main Service Channels FIC and MSC. For this purpose, the puncturing is performed with the aid of perforation vectors with coordinates w_i ($i = 0, ..., 31$) from the set $\{0,1\}$. Components of value 1 in each subgroup define the bits to be transmitted, see Fig. 2.28. Since each subgroup is generated by the 8 input vectors a (8 times 4 bits), the number of 1th in puncturing vector must be greater than 8 (input bit "a" must imply at least one outer bit x). The number of ones above 8 defines the puncturing index PI.

$$\sum_{i=0}^{31} w_i = 8 + PI$$

The puncturing index determines relation of the number of bits coming out of the encoder to the total number of bits in each subgroup:

$$(8 + PI)/32$$

and efficiency of the code for the subgroup:

$$8/(8 + PI)$$

With the increase of the puncturing index value from one (nine bits equal one in perforation vector) to 24 (all bits are equal one), the code efficiency for the subgroup steps through the set of fractions:

$$8/9, 8/10, 8/11, 8/12, \ldots, 8/30, 8/31, 8/32.$$

The code efficiency is thus contained within the range $\frac{1}{4}$, ..., 8/9.

Because of the perforation procedure of each subgroup with the perforation vector of fixed index, one gets perforated codeword of the group with a total length of $4(8 + PI)$.

Bits of "tail" are punctured with perforation vector of 24 bits (6 input zeros x 4 outer bits) which leaves from each vector x only the first 2 bits (Fig. 2.29).

In the communication system, the encoding subsystem is placed in front of the modulator in the transmitter and in receiver after demodulator.

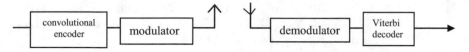

Fig. 2.29   Place of encoder in a communication system

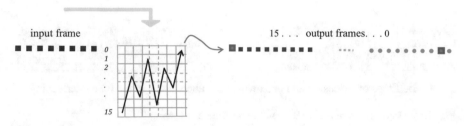

Fig. 2.30   The concept of time interleaving

More information on grouping codewords in the DAB channels is contained in the specification of the system [1].

2.6.3   Time Interleaving of the Frames of Encoder

In each of the subchannels of the DAB system, the output codewords of the convolutional encoder are subject to a time interleaving. The aim of time interleaving is a time separation of bits of data frame between sequence of outer frames.

Thanks to time interleaving, the interferences during transmission spoil only single bits of data frames, what can be improved in a convolutional decoder in receiver.

As shown in Fig. 2.30, the incoming frames are pushed, frame by frame, from top to bottom of the interleaver box. Every outgoing frame consists of nodes of a broken line. In every row and every column of the interleaving box, there is only one node. As a result, the bits of incoming frame will be separated between number of frames equal to the depth of the box. Thanks to this, the short-term disruption of the transmission channel, such as crackling, lightning, or dropouts, will not result in the loss of the entire codeword. The loss of single bits or their distortions may be supplemented or reconstructed with a high probability applying the Viterbi decoder in the receiver.

Time interleaving is characterized by the length of the incoming codewords and the depth of the encoder. In the DAB system, the depth of encoder is equal to 16 words [1], and frame length depends on efficiency of convolutional code assumed for a given subchannel. Parameters of the source coder and the convolutional encoder efficiency ensure that it is always a multiple of 16. This allows to repeat the structure of the output words of encoder (logical frames) starting every

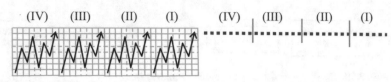

A. Steps of time interleaving of the frames having a length of 4 blocks

B. Steps of extension and narrowing the subchannels in the time interleaving

Fig. 2.31 Organization of the DAB frames in the time interleaving

16 columns, as shown in Fig. 2.31a. The blocks of 16 columns are repeated. The output words thus contain one bit from each column of each block and from each of the 16 rows.

The length of output words is equal to the length of the input words. This happens when all input words are of equal length.

However, if during operation of the encoder occurs simultaneously or separately:

- Reconfiguration of subchannels
- Change in the efficiency of convolutional encoder

the length of the input words is changing. Any change in the length of words results in a multiple of 16 bits, which is the full number of blocks. For these changes, there is selected an appropriate strategy of construction the output frames.

Extension of codewords at the input of interleaver causes extending the output words. The first 15 frames include new blocks with the words completed with zeros, as in Fig. 2.31b left.

Narrowing the codewords will not cause immediate changes in output words of encoder. Elements of the blocks designed to withdraw will be replaced with zeros during this period as in Fig. 2.31b on the right.

A consequence of assumed algorithm is operation of the DAB system during reconfiguration of the logical subchannels. Each proposed change requiring the narrowing or the extension of the subchannel must be announced for the receiver in advance of at least 16 frames.

2.7 Role of Multiplexer: DAB Logical Frames

The multiplexers play a main role in the generation of the DAB+ signal. Its function is to combine frames of the individual programs and data in the full logical frame of the DAB channel. The idea of a multiplexer is presented in Fig. 2.32.

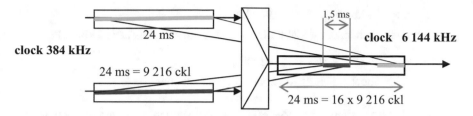

Fig. 2.32 Parameters of logic frames at input and output of the multiplexer

The throughput of individual audio subchannels is determined by the parameters of audio coders. The maximal throughput of the audio coder is 384 kbit/sec, what corresponds to 384 kHz clock rate. The duration of all output frames of encoder is equal 24 ms, so the maximum capacity of individual audio frame is 9216 bits.

In order to link, i.e., to multiplex, several of these subchannels, the clock speed at the output side must be a multiple of 384 kHz. The standard clock of this property is the standard T2 (USA) where second multiple has the speed 6144 kHz = 16 × 384 kHz. Application of the timer T2 theoretically allows to multiplex 16 different programs.

The theoretical throughput of such multiplexer is limited by the throughput of the following module of the system, i.e., the OFDM channel modulator, which is modulated in real time by a bit stream from multiplexer.

As explained in Sect. 2.8 the throughput of OFDM modulator is determined by system parameters:

- The frequency block of the size adopted for DAB (1536 kHz)
- Organization of subcarriers in the channel modulator
- Length of the OFDM modulated symbols
- Modulation system of subcarriers (D-4PSK)

Consequently, the throughput of the main service channel (MSC) is equal to

$$2\,304 \text{ kbit/sec} = 6 \times 384 \text{ kbit/sec}$$

regardless of the modulator mode. So the throughput of the DAB system allows to attach to multiplexer only six outputs of audio codecs or data frames with outer throughput 384 kbit/sec.

Using a higher compression ratio of audio codecs (DAB+), the other divisions of the main transmission channel throughput can be used. For example:

6×192 kbit/s plus 3×384 kbit/s

(six subchannels at 192 kbit/s and three subchannels at 384 kbit/s)

6×256 kbit/s plus 2×384 kbit/s

(six subchannels at 256 kbit/s and two subchannels at 384 kbit/s)

6 × 224 kbit/s plus 5 × 192 kbit/s

(six subchannels at 224 kbit/s and five subchannels at 192 kbit/s)

9 × 128 kbit/s plus 6 × 192 kbit/s

(nine subchannels at 128 kbit/s and six subchannels at 192 kbit/s)

In the DAB+ system, with more effective audio coder, necessary throughputs are nearly twice lower, and so the number of allowed subchannels and its combinations is adequately higher.

These divisions are concerning the gross throughput of individual subchannels, including the redundancy of the convolutional code, the CRC code, and the codes of headers.

In the transmitter of the DAB or DAB+ system, there are multiplexers:

1. The multiplexer combining audio frames of different subchannels and service packages in the main service channel (MSC).
2. The multiplexer combining Fast Information Channel (FIC) and the main service channel (MSC).
3. The multiplexer attaching Synchronization Channel (SC) to the channels FIC and MSC.

The functions of each multiplexer are illustrated in Fig. 2.33.

At the output of the multiplexer (1) there are formatted the logical frames of the main transmission channel MSC. The MSC channel throughput is determined by the throughput of the channel encoder OFDM discussed in the next section.

The combination of the output of the multiplexer with channel encoder requires an interface. In the DAB system, it is the interface WG1/WG2 organized as in Fig. 2.34 [10, Annex C].

The frames at the buffer input of 24 ms length count up to 9216 bits each for the maximum outer throughput. This corresponds to the input clock rate of 384 kHz.

Outputs of the audio codecs are included in sequence, every 9216 cycles of multiplexer clock.

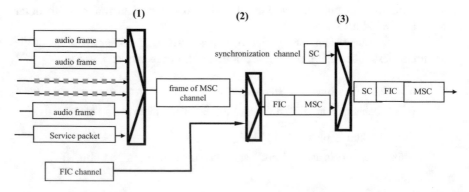

Fig. 2.33 Building a logical frame of the DAB system

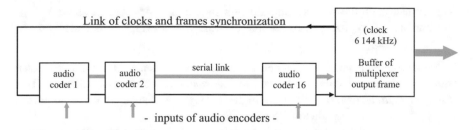

Fig. 2.34 The concept of interface WG1/WG2 (after [10, Annex C])

One frame of the multiplexer output buffer contains 147,456 cycles with a duration of 24 milliseconds. This is due to the output clock rate equal to 6144 MHz.

The capacity of the output buffer of the multiplexer is equal to 16 capacities of the input frames of audio coders in case of maximum capacity. Because of the limited capacity of the following subsystem – OFDM modulator – only a part of the signal is used to modulate the OFDM coder. So, the buffer will contain several "blank spaces," which are not further transmitted.

In addition to audio encoders, the multiplexer can connect any service organized in 24 ms frames with a capacity not higher than 9216 bits.

Each logical frame in the DAB(+) system consists of a synchronization field, Fast Information Channel field, and a main service channel field, as mentioned in Sect. 2.3 (Fig. 2.5).

The multiplexer (2) in Fig. 2.33 links the frames of the MSC channel with frames of the FIC channel. This function of the multiplexer can be realized through joint buffering both frames and next sequentially further processing.

The synchronization field and field of the initial phases is connected via the multiplexer (3) to the frames of the FIC and MSC channels.

As a result of operations of multiplexers (1), (2), and (3) of Fig. 2.33, the fields of logical subchannels are combined into logical DAB(+) frames. This process is related with parameters of the OFDM modulator.

A modern viewpoint on the role of multiplexers is presented in webinars [11, 12].

Subchannels are rented by the service providers. Their demands are defined by bit rates of transmitted applications. The throughput of the DAB channel is limited by throughput of the OFDM encoder, so it follows that the possible extension of the subchannel of one provider is associated with limiting the throughput of another provider. Changing in time the subchannel throughputs requires clear rules. As a reminder:

- Reduction of the capacity of logical subchannel frame does not cause immediate reduction of the subchannel throughput.
- Time of transition to a lower throughput of subchannel requires 384 ms (16 logical frames).

– Information on the subchannel reduction should precede this fact about 384 ms
 (time necessary for reorganization of time interleaving in decoder on the
 receiver side).

Such problem appears, e.g., when increasing the efficiency of the convolutional
code for subchannel data, as it leads to increasing capacity of subchannel. This
results in an immediate transition to a higher throughput of subchannel, although the
first 15 frames will be padded with zero bits.

Hence temporal correlation during reorganization of subchannels: narrowing and
expanding. This relationship must be considered in the process of reconfiguring the
multiplexer.

2.8 OFDM Channel Modulator

The received DAB signal can be distorted in amplitude and phase. These nonlinear
distortions can be reduced to linear changes – what facilitates demodulation – if the
baseband frequency block of the DAB signal is split into narrow subchannels with
their own subcarriers.

Modulation of the DAB channel is carried out in a specially developed system
where subcarriers within frequency block B are independently modulated. In transmitter the individually modulated subcarriers are summed up forming a DAB
symbol. The subcarriers are formed as an orthogonal system of sinusoids which
allows the use of fast Fourier transform (FFT) in a demodulating process. This
system of channel encoder is called the Orthogonal Frequency Division
Multiplexing (OFDM) channel modulator.

The idea of the OFDM modulator is shown in Fig. 2.35. On one axis, there is
time, on the other, frequencies of subcarriers (it is not the full frequency axis) within
the block B, and the vertical axis represents adequate amplitudes of subcarriers.

The signal is divided into time slots of duration T_U. This gives rise to the OFDM
symbols. Each OFDM symbol consists of N subcarriers with frequencies mutually

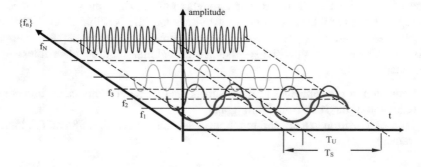

Fig. 2.35 The physical image of subcarriers of the OFDM channel modulator

orthogonal. Each subcarrier is modulated independently. The sum of modulated subcarriers within time interval T_U creates an output symbol of the OFDM modulator.

For simplicity of presentation (identifying subcarriers and amplitude axes), the OFDM signal is displayed graphically as in Fig. 2.36.

Requirement of orthogonality of the subcarriers resulting Fourier transform in the time interval $<0 - T_U>$ is guaranteed by the selection of the N subcarrier frequencies:

$$f_k = k \left(1/T_U \right) \text{ for } k = 0, 1, \ldots, N - 1.$$

The frequency characteristic of time limited segment of sine function with frequency f_k is described as

$$\frac{\sin \left(2\pi \{f - f_k\} \right)}{2\pi (f - f_k)},$$

so, a total of characteristics of the OFDM symbol is the sum of the following sliding charts for k = 0, 1, ..., (N−1). The frequency characteristic of the OFDM symbol in the baseband channel appears as envelope of the sum of characteristics of individual subcarriers (Fig. 2.37).

Fig. 2.36 An image of subcarriers of the OFDM symbol

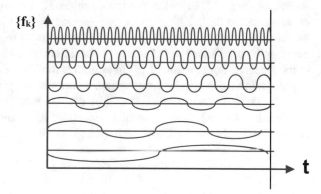

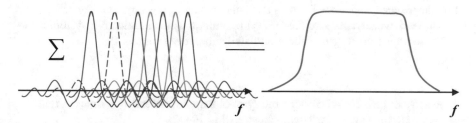

Fig. 2.37 The frequency envelope of the OFDM symbol

2.8.1   The Principle of Operation

The OFDM modulator or channel encoder – both names are used – has been introduced for the first time in the DAB system [13], included into EU norm, and further developed in, e.g., [14].

2.8.1.1   Why Modulator with Multiple Subcarriers

The maximum throughput R of a bit stream transmitted via frequency block B with a modulated single carrier according to the Shannon theorem is equal to

$$R = B \cdot \ln_2 M,$$

where M – number of modulation points

The duration of symbols in the single-carrier system (system with single-frequency carrier is also referred to as single-tone as opposed to multitone encoder system using multiple subcarriers):

$$T_s \, [s] = 1/B \, [Hz]$$

In a block of 1.5 MHz the symbol duration T_s is equal to 0.66 μs. Meanwhile, experiments on the propagation of radio waves in the VHF band indicate that the signal delay due to multipath propagation is up to 20 microseconds. Thus, due to overlapping of reflected signals, there occurs effect of inter-symbol interference (ISI) in the receiver. Suppression of ISI interference due to multipath propagation would require in the receiver echo cancellation, a complex adaptive structure.

If the same frequency block B will be divided into N subchannels, each with a bandwidth B/N and with individual subcarrier modulated by the same M-point scheme – a total capacity R_N of such system will be:

$$R_N = N \cdot (B/N) \cdot \ln_2 M = R$$

i.e., the same as in the single carrier system.

Regardless of the number of subcarriers in the OFDM encoder, the throughput in a predetermined frequency block is the same as that of the single carrier case.

The decisive advantage of the OFDM modulator as the multicarrier system is the ability to adjust the length of the modulated subcarrier symbols T_N:

$$T_N = 1/(B_N) = 1/(B/N) = N \cdot T_S$$

In a system with N subcarriers, the symbol length is extended N times in relation to the single-tone symbols. This is illustrated in Fig. 2.38.

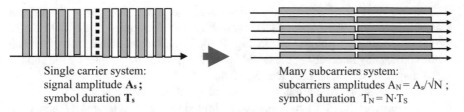

Single carrier system: Many subcarriers system:
signal amplitude A_s;   subcarriers amplitudes $A_N = A_s/\sqrt{N}$;
symbol duration T_s symbol duration $T_N = N \cdot T_S$

Fig. 2.38 Relations between single carrier and many subcarrier systems

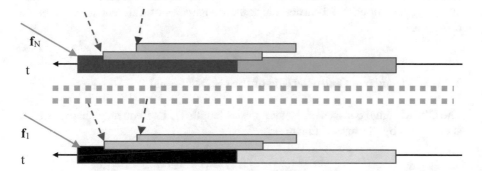

Fig. 2.39 The influence of the ISI interference on the state of the multicarrier encoder

Increasing the number of subcarriers N in the block B the length of the OFDM symbol can be extended so that the interference caused by the multipath influences only a few symbols, as in Fig. 2.39.

This is important because the echo cancellation circuit in the receiver of such system is greatly simplified. Of course, it requires the number of subcarriers leading to symbols length elongation not less than the maximum multipath delay. For our example (block of 1.5 MHz, the maximum delay time of 20 μs, the duration of the symbols for one carrier system 0.66 ms), the minimum N satisfying this condition is $N_{min} = 20\ \mu s/0.66\ \mu s = 30$.

The number of subcarriers in a DAB+ transmitter of mode I is equal to 1536, so the condition of limiting the inter-symbol interference is filled up.

2.8.1.2   Introduction of the Guard Intervals

The signal processing in the DAB+ receiver can be further simplified by introducing between the OFDM symbols, the so-called guard intervals of length T_g greater than maximum time delay τ_M. As shown in Fig. 2.40, if duration of the guard interval is greater than the maximum time delay of reflections, signals delayed by multipath transmission will not interfere with adjacent following symbols. Thanks to the introduction of the guard interval, the inter-symbol interference problem (ISI) in the multitone encoder disappears completely.

The introduction of an empty guard interval solves the problem of the ISI but leaves intercarrier interference (ICI), caused by a discontinuity in the field T_U of signal processing, as pointed out in Fig. 2.40. To ensure continuity of the signal within processing interval T_U, the guard interval is filled in transmitter by the end the OFDM symbol signal, as shown in Fig. 2.41.

The part of signal filling the guard interval is called a cyclic prefix.

In the receiver the FFT demodulation is limited to useful (orthogonal) field T_u of length equal to the inverse of the inter-tone spacing $T_u = 1/\Delta f$. The use of the Fourier transform for symbols demodulation allows in this case for calculating the modulating symbols of each subcarrier, since the frequencies of different subcarriers are orthogonal.

2.8.1.3 Frequency Characteristic of the OFDM Signal

The OFDM signal consists of the symbols of length T_u. Its frequency characteristics is described by the sum of functions:

$$\sin\ [2\pi(f - f_k)T_U]/2\pi(f - f_k)T_U$$

with maxima at the points $k \cdot (1/T_U)$ and a flat envelope of Fig. 2.37, which guarantees the optimal use of the block spectrum. The first zero of each single function, next to the coordinate of the maximum, is located at $1/(2T_U)$ from the coordinate of maximum.

Introduction of the guard intervals Tg between useful fields of OFDM symbols, filled with the cyclic prefix, does not change the frequencies of subcarriers, still equal to $k \cdot (1/T_U)$, thus leaving the spacing between subcarriers. In contrast, the frequency characteristics of the subcarriers is broadening by Tg giving new functions

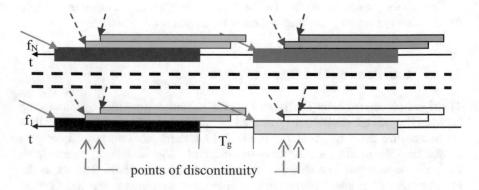

Fig. 2.40 Elimination of inter-symbol interference by introducing a guard interval

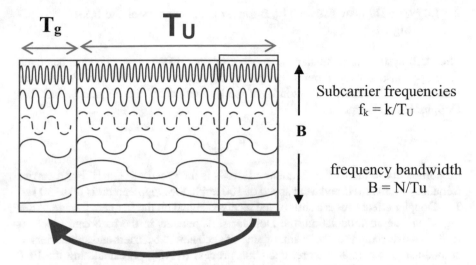

Fig. 2.41 Filling the guard interval with the cyclic prefix in the OFDM symbol

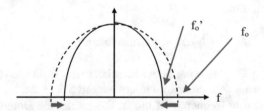

Changing the frequency envelope of a single sub-carrier when including the guard interval

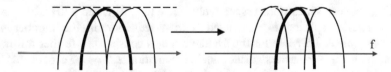

Fig. 2.42 Change of the envelope of the OFDM signal after introduction of the guard intervals

$$\sin \left[2\pi(f\text{-}f_k)(T_U + T_g)\right]/2\pi(f\text{-}f_k) \, (T_U + T_g)$$

In this case, the first zeros of each function will be within $1/\{2(T_U + T_g)\}$ of the coordinates of maxima – the individual frequency characteristics will be narrowed compared to the previous characteristics. It manifests in the occurrence of minima between the maxima of the envelope of total signal, as shown in Fig. 2.42.

2.8.1.4   The Doppler Effect: The Impact on Parameters of the DAB Signal

The DAB system, in particular the concept of the OFDM channel coder, has been developed also for the moving reception, e.g., when driving a car. For the car velocity v_k with respect to the source of the k-to reflection, the frequency shift (the Doppler effect) is equal to

$$f^D_{\ k} = F_c \cdot v_k/c$$

proportionally to the carrier frequency F_c (c – is the speed of signal). For a carrier frequency of 300 MHz and a car speed of 100 km/h, the Doppler shift is up to 30 Hz. The Doppler effect causes a shift of the received signal on the frequency scale. After high-frequency demodulation, the Doppler shift remains in the baseband block. It leads to the deviations of the orthogonality of such new subcarriers and subcarriers in demodulation block. Therefore, decoding process (relying on calculating the FFT coefficients of the OFDM symbol) will introduce errors of modulating symbols of received subcarriers.

In order to decode the deviated symbols with acceptable probability, it is necessary to fulfill the condition

$$f^D_{\ k}/\Delta f < \text{permissible error}$$

where Δf is the frequency spacing between subcarriers. In the DAB+ system, it is assumed that the permissible error should not exceed 5% of Δf.

The higher the carrier frequency of block, the greater the Doppler shift and the greater the possible errors of the OFDM decoder. To reduce the error for higher carrier frequency, it is necessary to increase the spacing between the subcarriers. This was the reason of introducing in DAB system the different modes of the codecs with spacing between subcarriers, respectively, equal to 1, 2, 4, and 8 kHz. Increasing the intercarrier spacing is associated with a reduction in the number of subcarriers (the block 1.5 MHz for the system remains constant). It further follows that maintaining the subcarriers' orthogonality condition $\Delta f = 1/Tu$ demands a reduction of the length of useful field of symbols and the length of the guard interval. However, as known from experimental data, for the higher carrier frequency, the delays of the significant signal reflections are smaller. So, the guard interval for the higher frequency bands can also be shortened.

In system DAB+ with carrier frequency limited to range 174–300 MHz, the subcarrier spacing was assumed to be 1 kHz (mode I).

2.8.2   Implementation of the Digital OFDM Modulator

With the digital signal processing technique, the output signal of the OFDM encoder is implemented without generating the physical subcarriers. Generating 1.5 thousand subcarrier frequencies equally spaced with the accuracy required for the OFDM modulator with their simultaneous modulation is practically not possible in the analog technology. Thanks to the technique of the digital signal processing, implementation of the encoder has become a reality.

2.8.2.1   Sampling the OFDM Signal

The digital technology allows to calculate the values of the baseband OFDM symbol with the sampling frequency at least twice the maximum OFDM frequency block. It is calculated independently for the quadrature and the in-phase symbol components $S(t_n) = S(n{\cdot}\Delta t)$.

$$S(t_n) = \sum_{k/0}^{N} e^{j\varphi_k} \cdot e^{j\omega_k \cdot t_n} = IFFT\left\{ e^{j\varphi_k^{(l)}} \right\}$$

Here the modulated differential phase of k-th subcarrier $\varphi_k^{(l)}$ in the l-th symbol is equal to the previous symbol phase $\varphi_k^{(l-1)}$ altered by the modulating symbol $y_k^{(l)}$.

$$\varphi_k^{(l)} = y_k^{(l)} + \varphi_k^{(l-1)}$$

Calculation of the OFDM signal at intervals $n{\cdot}\Delta t$ is thus reduced to the calculation of the discrete inverse Fourier transform (IFFT) of the known modulating symbols. It is performed in specialized, high-speed large-integration chip and must be fast enough to secure the modulation in real time.

Block diagram of the modulator is shown in Fig. 2.43:

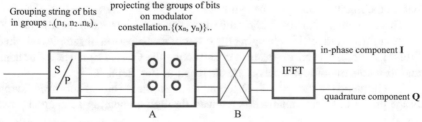

S/P - serial / parallel converter
A - constellation of encoder modulation: string of bits $(n_1, n_2, ...,n_k)$ → modulating vectors $\{(x_n, y_n.)\}$
B - frequency interleaver
IFFT – inverse Fourier transform

Fig. 2.43 The flowchart of the OFDM modulator

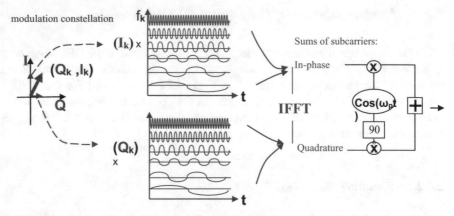

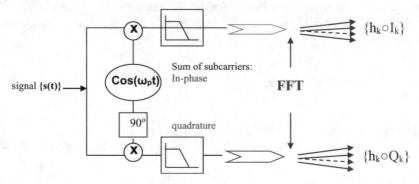

At the receiver, the signal processing occurs in the opposite direction:

Fig. 2.44 The stages of the OFDM signal processing in a transmitter and receiver

The phase and quadrature components, after the digital-analog converter, are placed on the input of an intermediate modulator with frequency ω_p, from where the signal – through the HF system – is emitted into the ether.

In order to install the guard interval with cyclic prefix into the signal, the samples of the final part of the OFDM symbol within the time segment Tg are copied and put before the initial samples of the symbol.

This implementation of cyclic prefix into the guard interval in the digital OFDM encoder is thus reduced to changing the order of transmission of samples within each OFDM symbol. The basic stages of constructing the OFDM signal at the transmitter and its reconstructing in receiver are sketched in Fig. 2.44.

The influence of the channel observed as multipliers $\{h_k\}$ of subcarrier symbols, in the process of demodulation, is reduced through comparing actual and previous values:

$$\left(h_k^{(l)} \cdot I_k^{(l)} / h_k^{(l-1)} \cdot I_k^{(l-1)}\right) \approx \left(I_k^{(l)} / I_k^{(l-1)}\right)$$

It happens because the changes of channel between successive symbols are small enough:

$$h_k^{(l)} \approx h_k^{(l-1)}$$

The phase difference between $I_k^{(l)}$ and $I_k^{(l-1)}$ is thus equal to the phase of the modulating symbol of the k-th subcarrier.

$$\varphi_k^{(l)} - \varphi_k^{(i-1)} = y_k^{(l)}$$

2.8.3   Transforming the DAB Logical Frames into Modulating Vectors

Stages of transforming groups of bits of the logical DAB frame into the complex factors modulating OFDM subcarriers of the physical frame are illustrated in Fig. 2.50.

Every logical DAB frame is gradually divided into groups P_l counting 2 N bits, where N is the number of subcarriers in the OFDM modulator. Bits of groups are subsequently converted into the complex numbers and the modulating factors through the following steps.

A. Gray Mapping

For every couple of bits $(p_k, p_{(k + N)})$ for $k = 1,..,N$, each modulating vector P_l of Fig. 2.50 is assigned the complex number q_k according to the formula

$$q_k = \left\{ (1\text{-}2p_k) + j\left(1\text{-}2p_{(k+N)}\right) \right\}/2$$

The phase of q_k identifies unambiguously the pair of bits $(p_k, p_{(k + N)})$. In the real signals formulation, instead of complex numbers, the in-phase and quadrature components of the signal are used.

B. Frequency Interleaving

The purpose of interleaving is to place consecutive modulating symbols q_k between separated subcarriers. As a result, an interference in the frequency domain distorting neighboring subcarriers within some range does not destroy modulation symbols representing adjacent bits of one logical frame.

Frequency interleaving [1] is operation of permutation of complex numbers q_k modulating the subcarriers of the OFDM symbol. The number of permuted complex symbols is equal to the number of subcarriers of the OFDM encoder. For an assigned mode of the DAB system, the same permutation operation is applied in each OFDM symbol (see Fig. 2.50).

$$\{y_k\} = \text{permutation}\{q_k\}, \quad k = 1, 2, \dots, N$$

C. Phase-Differential Modulation

In the DAB radio, the differential phase shift keying (D-4PSK) modulation is applied. It requires a fixed initial reference phase, so-called reference phase, for each subcarrier at the beginning of each DAB frame in the synchronization symbol. The decisive advantage of the phase-differential modulation is its increased resistance to distortions during multipath propagation.

The reference phase symbols at the first symbol of each OFDM frame belong to the set of angles 0, $\pi/2$, π, $(\frac{3}{4})\pi$. Specific values for each subcarrier are given in the norm of the system in the form of Table [1].

In transmitter the reference phases are used to initiate the process of demodulation: calculating the phases between consecutive symbols:

$$\varphi_k^{(l+1)} = \varphi_k^{(l)} + y^{(l+1)}_k$$

At the receiver, due to a signal distortion during transmission, a decoded symbol with its phase is generally different from the transmitted one, i.e., $\varphi'_k(1) = / = \varphi_k(1)$. But the following symbol is similarly distorted, so the difference is alike in the transmitter.

A phase differential modulation system allows with high probability for a correct demodulation. More on this topic, see Chap. 3 ("The Impact of Channel Propagation on the Quality of DAB Signal Reception").

2.8.4 Peak to Average Power Ratio (PAPR): Energy Dissipation

The frequency envelope of the output signal in transmitter is shaped by the channel OFDM modulator and an output filter. To hold signal within the linear characteristic of the output power amplifier – the peak to average power ratio should be small enough. The peaks in the OFDM signal appear in the points of overlapping maximas of subcarriers (see Fig. 2.45a). To avoid such situation, the phases of the subcarriers should be close to random. This means that the groups of bits mapped on the modulation constellation should have a distribution of zeros and ones close to random. However, in the real signal, there can appear strings of zeros and ones with distribution different from random. Similar effects occur during transmission of data services. In the encoder, such effects lead to the accumulation of the maxima of subcarriers.

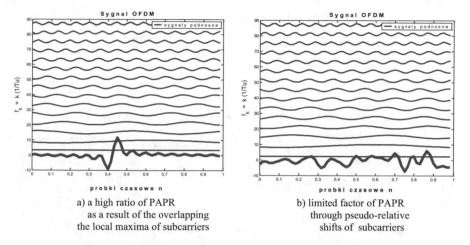

a) a high ratio of PAPR
as a result of the overlapping
the local maxima of subcarriers

b) limited factor of PAPR
through pseudo-relative
shifts of subcarriers

Fig. 2.45 Illustration of the PAPR phenomenon in the OFDM system

The relation of peak to average power in the OFDM symbol, referred to by the acronym PAPR (Peak to Average Power Ratio), where the average power is the rms value (root mean square: $rms = \{\Sigma\ p_k \cdot S_k^2\}^{1/2}$), has the essential meaning for the OFDM signal transmission. The time course of the baseband OFDM signal is the basis of the received signal demodulation, so the fidelity of its waveform determines the reliability of the transmission. Signal pulses significantly higher than the mean value require the use of power amplifier with linear characteristic in a large range of signal amplitudes, which is technically difficult to implement. For this reason, the controlled differentiation of the phase-modulated subcarriers in the outer OFDM signal is proposed [1] to avoid high PAPR ratio. Impact of differentiation phases on reduction of the coefficient of PAPR is illustrated in Fig. 2.45b.

Not adjusted high PAPR factor forces limiting the power level of the output signal and so the range of coverage area of the DAB transmitter.

The upper limit on the PAPR ratio in the output signal of terrestrial DAB(+) transmitters is 13 dB.

The binary groups of the real signal will take a random form if in a controlled manner it will be added (modulo two) to the pseudo-random sequence. The receiver will recover the actual string again adding to the resulting synchronized signal the same pseudo-random sequence.

In the DAB system, where the processed bit strings have hundreds of bits, the pseudo-random string is generated by a shift register.

The initial word of register is created by ones. Synchronization of pseudo-random generators in the transmitter and receiver is achieved by synchronizing its operation with a fixed beginning.

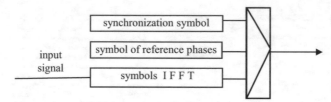

The resulting frame of N symbols of the OFDM modulator has the form:

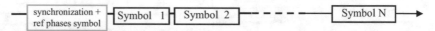

Fig. 2.46 Forming the OFDM signal frame

2.8.5   *Output Physical Frame of the OFDM Modulator*

The physical OFDM frame consists of individually modulated symbols. The number of symbols in a frame depends on the mode of generator OFDM. In the final multiplexer the formatted frame of data symbols is supplied with the first two symbols of each OFDM frame: zero symbol synchronization (0) and phase reference symbol (1) (see Fig. 2.46).

2.8.6   *Mode of the OFDM Channel Encoder in the DAB+ System*

Parameters of the OFDM channel encoder (channel modulator) arise from the following assumptions:

- The width of the frequency block B [MHz]

As a result of the international arrangements, a grid of four-element frequency blocks was applied for conveniently pre-planning the frequency allocation for the DAB system: the 6 MHz TV channel was divided into four blocks, so a single block consists of 1.5 MHz. The resulting maximum sampling frequency of the signal appears $f_{max} = 1.5$ MHz, as the reciprocal of the probing step T, and the probing step is limited by the bandwidth, $T = 1/B$.

- The width of the frequency spacing between subcarriers Δf

To minimize the negative impact of Doppler shift on the processes of synchronization and demodulation in the receiver, it is assumed that the maximum Doppler shift should not exceed 5% of the distance between the subcarriers. Doppler shift depends on the maximum speed of terminal and the carrier frequency of the system. To fulfill the 5% condition for the DAB+ system blocks in the full range of band frequencies, the inter-subcarrier spacing Δf was assumed 1 kHz.

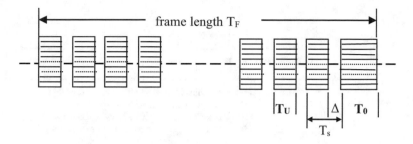

Fig. 2.47 Parameters of the physical OFDM frame

Table 2.4 Parameters of the channel encoder OFDM, mode I [I]		
Parameter	Mode I
OFDM frame T_F	96 ms
Zero symbol T_0	1.297 ms
Useful (orthogonal) field	1 ms
Guard interval Δ	0.246 ms
Full symbol T_s	1.246 ms
Number of subcarriers	1536
Interspace between subcarriers	1 kHz
Number of symbols in frame	77

In the frequency block B, the number of subcarriers N within block is obtained as a result of division B/Δf. Not all these subcarriers are actively used, since the shape of the mask of the DAB+ signal in the baseband channel limits the use of subcarriers at the edges of the block.

• The modulation/demodulation system based on the Fourier transform

Orthogonality of the subcarriers in Fourier sense is convenient for fast demodulation process in the receiver. It requires that the length of the useful field T_U of OFDM symbols is inversely proportional to the spacing between the subcarriers: $T_U - 1/\Delta f$. According to the assumed variant of subcarrier spacings, the specification of the system considers field lengths of the useful OFDM fields.

• The length of a guard interval between symbols

In the DAB+ system, there was adopted a length of guard interval equal to ¼ the length of the useful field (Fig. 2.47).

The meaning of individual parameters is illustrated in Fig. 2.50. Specific data of mode I are presented in Table 2.4.

Operating mode is related to the working conditions of the transmitter. Mode I provides the greatest resistance to interference due to longest intervals between OFDM symbols in a frame. For the same reason, it allows for a greater distance between the transmitters in the single-frequency network (see Sect. 4.2).

Graphical presentation of the OFDM frame parameters is indicated in Fig. 2.48.

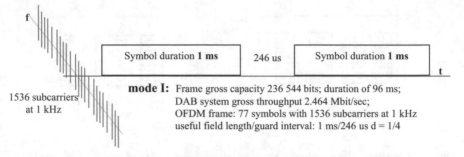

Fig. 2.48 Frame parameters of the OFDM encoder in the DAB system

2.9   Mapping the Logical DAB Frames onto the Physical OFDM Frames

Multiplexing the frames of individual programs and services into the Main Service Channel forms the MSC field of the DAB+ logical symbol. Organization of this field is described in the Fast Information Channel (FIC) composed of Fast Information Block fields placed before MSC. The logical DAB+ frame is finally formed after adjoining the Synchronization field (S) at the beginning.

To transmit data, the DAB+ logical frame must be mapped onto the physical carrier of information, i.e., the OFDM symbols forming the OFDM frame.

Generally, the mapping:

$$DAB + \text{logical frame} \rightarrow OFDM \text{ physical frame}$$

requires fulfillment of two conditions:

- The bit capacity of both frames must be equal.
- The duration of each frame must be the same.

The OFDM encoder parameters depend on the operation mode (see Table 2.4). To make operation of mapping independent of modes, the MSC logical frame is built of fundamental, mode independent blocks. The building blocks adopted for the DAB and DAB+ logical frames are the *Common Interleaved Frames (CIF)*. The name comes from the fact that the original frame of audio or services, which by time interleaving is scattered among other 16 frames, does not go beyond the CIF frame. As a result, at the receiver, the signal processing, leading to the restoration of the original audio or services frames, shall be limited to one CIF frame. The organization of each CIF is described in the Fast Information Blocks (FIB) in the Fast Information Channel (FIC), typically, three FIB for one CIF frame. The DAB logical frame consists of 4 CIF and 12 FIB fields.

The DAB+ logical frames are projected onto the OFDM symbols, which are carriers of the physical information. Schematically, this relationship is shown in Fig. 2.49.

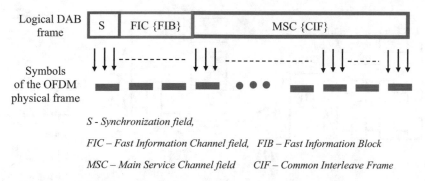

S - Synchronization field,

FIC – Fast Information Channel field, FIB – Fast Information Block

MSC – Main Service Channel field   CIF – Common Interleave Frame

Fig. 2.49 Mapping the logical DAB+ frame onto the symbols of the OFDM frame

Capacity of the OFDM frame results in the number of OFDM symbols and the capacity of one symbol. A single symbol carriers two bits per subcarrier so twice the number of subcarriers. This follows the adopted modulation Scheme D-4PSK.

Dividing the capacity of the CIF field by capacity of a single OFDM symbol, one obtains the number of OFDM symbols required for accommodating the content of the field CIF. Similarly with the FIB fields. When the number of sets (CIF + 3FIB) within field CIF is not natural, some parameters should have been respectively increased.

The time duration of the DAB+ logical frame including the synchronization field (zero and phase reference) must be equal to the duration of one OFDM frame based on the number of symbols, its length, and the guard interval adopted for selected mode.

Mapping the DAB+ logical frames onto the OFDM frames is shown schematically in the Fig. 2.50.

Mapping comprises the following steps:

- Splitting the CIF and the FIB fields onto the uniform "modulating" groups P_I, each with capacity of the one OFDM symbol
- Assigning the pair of bits in each group to one modulating factor q of the D-PSK modulation scheme
- Permutation of modulation factors according to frequency interleaving scheme constant in each symbol
- Modulating subcarriers of every OFDM symbol with the corresponding modulating factors

In Fig. 2.50, the mapping is obtained for the four building blocks (CIF + 3 FIB); hence, duration of the logical frame in this mode is $4 \times 24 = 96$ milliseconds.

The indicated mappings refer to the DAB+ frames including the redundancy elevated because of convolutional coding. In the Fast Information Channel (FIC), the constant code performance of 1:3 is adopted, hence the known net and gross capacity. In the main service channel (MSC), variety of content is encoded with different coefficients; hence, the only stable parameter is capacity of the fields CIF.

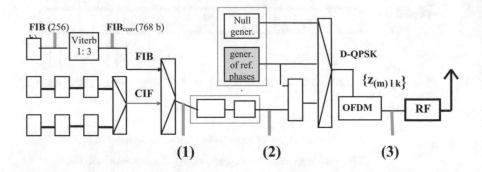

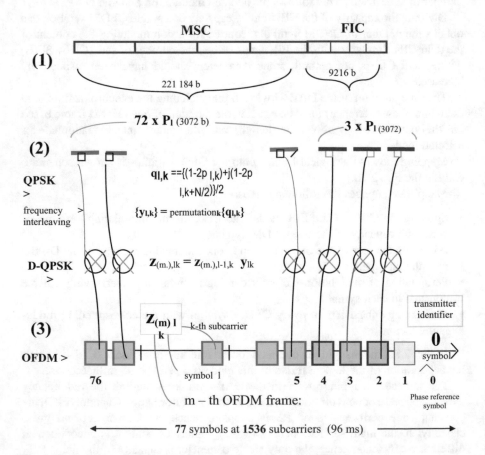

Logical DAB frame, **Mode I:**

MSC frame = 4 CIF fields; CIF field = 864 units CU (24 ms); 1 unit CU = 64 bits
frame FIC = 4 fields FIB

Fig. 2.50 Projecting the logical DAB frame onto the physical OFDM frame

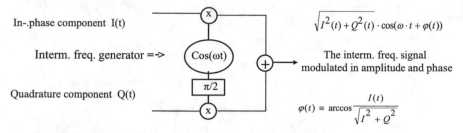

Fig. 2.51 Modulator 2-PSK

2.10   Integrated Transmitter Output: The Output Filter

The real and complex components (in-phase and quadrature) of digital signal from OFDM encoder are converted to analog signals in digital-to-analog converters (A/D). The two components are next converted in 2-PSK modulator into an intermediate signal of frequency $\omega = 2\pi f$, as shown in Fig. 2.51.

The following element of the DAB transmitter is a block HF.

2.10.1   The Output Frequency Mask

An important element of the DAB+ transmitter is HF output filter. Its task is signal attenuation to avoid interference of DAB+ with other systems operating in neighboring frequencies blocks. Based on simulations, the mask of outer filter attenuation was defined as a function of frequency near carrier frequency. Mask of DAB+ blocks for VHF band according to [1] illustrates the graph in Fig. 2.52.

2.11   Functional Description of the DAB Receiver

In the receiver (terminal), the high-frequency (HF) DAB+ signal is transferred to a baseband block and then processed in the order reverse to that in the transmitter. Digital signal can be processed in hardware or software. For this reason, it is purposeful to describe receiver functions as functional blocks instead of a detailed description of the specific solution.

Organization of services listed in receiver profiles depends on the optional software and is left to the decision of producers.

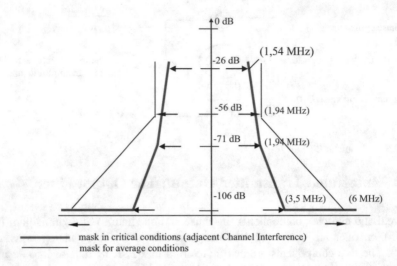

Fig. 2.52 Block of the DAB mask in the VHF frequency band [1] (The thin line is applied in case of empty adjacent blocks)

2.11.1   Block Diagram of the DAB Receiver

The overall functional block diagram of the receiver is presented in Fig. 2.53.

In the DAB+ receiver, the physical OFDM frames are transformed into the DAB+ logical frames:

- After antenna, the high-frequency block converts HF signal into baseband block.
- After time and frequency synchronization, CP extraction and 2-PSK demodulator recover I and Q components.
- The FFT transform reconstructs distorted individual OFDM factors.
- Demodulation D-4PSK on individual subcarriers reconstructs modulating factors.
- Frequency deinterleaving and inverse modulation constellation allow to recover sequences of coded bits.
- Time deinterleaving within FIC frames renders bits in the ordered convolution sequences.
- Viterbi decoder recovers original bit sequences within frame.

Details of the construction of receivers depend on the manufacturers: design of equipment, functions of individual blocks in integrated circuits, or the level of applied software solutions.

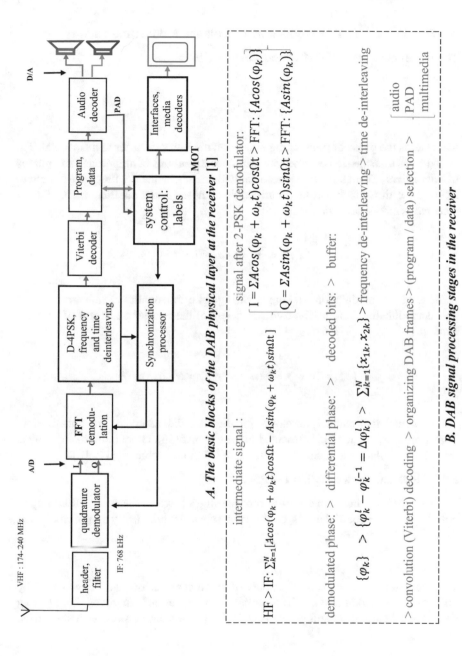

VHF : 174-240 MHz
header, filter
IF: 768 kHz
quadrature demodulator
A/D
I
Q
FFT demodu-lation
D-4PSK, frequency and time deinterleaving
Synchronization processor
Viterbi decoder
Program, data
system control: labels
Audio decoder
PAD
D/A
MOT
Interfaces, media decoders
A. The basic blocks of the DAB physical layer at the receiver [1]
intermediate signal :
signal after 2-PSK demodulator:
HF > IF: $\sum_{k=1}^{N}[Acos(\varphi_k + \omega_k t)cos\Omega t - Asin(\varphi_k + \omega_k t)sin\Omega t]$
$I = \Sigma Acos(\varphi_k + \omega_k t)cos\Omega t$ > FFT: $\{Acos(\varphi_k)\}$
$Q = \Sigma Asin(\varphi_k + \omega_k t)sin\Omega t$ > FFT: $\{Asin(\varphi_k)\}$
demodulated phase: > differential phase: > decoded bits: > buffer:
$\{\varphi_k\}$ > $\{\varphi_k^l - \varphi_k^{l-1} = \Delta\varphi_k^l\}$ > $\sum_{k=1}^{N}\{x_{1k}, x_{2k}\}$ > frequency de-interleaving > time de-interleaving
> convolution (Viterbi) decoding > organizing DAB frames > (program / data) selection >
audio
PAD
multimedia
B. DAB signal processing stages in the receiver

Fig. 2.53 Block diagram of a DAB receiver

2.11.2   Recovering the Logical Frames from the OFDM Signal

A. The OFDM Signal Demodulator: Calculating Modulating Factors

The phase-modulated OFDM symbols have the form

$$S(t) = \sum_{n=0}^{N-1} e^{j\varphi_k} \cdot e^{j\omega_k t},$$

where φ_k is the phase of modulating phasor on the k-th subcarrier and $\omega_k = 2\pi k/Tu$.

Wireless transmission channel can introduce distortions of amplitudes and phases of subcarriers. Thanks to division on individual subchannels with quasi-flat envelopes, these distortions can be described as complex linear coefficients h_k. So the received symbols have the form

$$S_h(t) = \sum_{k=0}^{N-1} h_k \cdot e^{j\varphi_k} \cdot e^{j\omega_k t}$$

After FFT demodulation – taking advantage of orthogonality of subcarriers – one gets demodulating factors. The p-th coefficient of the Fourier transform of set S_h is equal to

$$(p) = 1/N \sum_{n/0}^{N-1} S_h(n) \cdot e^{-j\frac{2\pi}{N}p \cdot n} = N \cdot h_p \cdot e^{j\varphi_p}$$

The actual minimal requirements for digital radio receivers are contained in Technical Specification [15] presented in the webinar [16]. Other modern viewpoints on today's digital radio technology are included in the webinars [17, 18].

B. Differential Demodulation D-4PSK

Experiments indicate that during broadcasting of two consecutive OFDM symbols, the complex coefficients h_k on the same subcarrier practically do not change, so

$$h_k^{(l)} \cong h_k^{(l-1)}$$

Depending on this equality the phase of modulating factors can be recovered by differential demodulation. Differential demodulation involves comparing demodulated current (*l*) and previous (*l-1*) coefficients on the same k-th subcarrier, i.e.,

$$\frac{h_k^{(l)} \cdot e^{j\varphi_k^{(l)}}}{h_k^{(l-1)} \cdot e^{j\varphi_k^{(l-1)}}} = e^{j \cdot \left[\varphi_k^{(l)} - \varphi_k^{(l-1)}\right]}$$

But the phase difference of adjacent coefficients is equal to the phase of modulating factor on the k-th subcarrier.

An important and still difficult problem in DAB reception is the simultaneous decoding of all OFDM subcarriers. To receive only the selected program, such a requirement is not necessary, because it is enough to know the factors associated with the selected program. However, to make full use of the DAB throughput, it is necessary to identify and decode also services non-associated with the program.

The bottleneck of this solution is time-consuming implementation of Viterbi algorithm for all the symbols in real time.

The initiation of the process of demodulating each successive OFDM frame requires knowledge of the initial phase references. They are included in the synchronization OFDM symbol of the DAB frame.

After recovering logical frames follows the choice of the selected program in the receiver.

Algorithm presenting startup procedures is described in Sect. 6.5.

References

1. ETSI EN 300 401 „Radio broadcast systems: Digital Audio Broadcasting (DAB) to mobile, portable and fixed receivers"
2. ETSI TS 103 466 "Digital Audio Broadcasting (DAB); DAB audio coding (MPEG LAYER II)"
3. ISO/IEC 11172–3. "*Coding of Moving pictures and associated audio for digital storage media at up to 1.5 Mbit/s - Audio Part*". International Standard, 1993
4. ISO/IEC 13 818–3. "*Information Technology: Generic coding of Moving pictures and associated audio - Audio Part*". International Standard
5. ETSI TS 102 563" Digital Audio Broadcasting (DAB); Transport of Advanced Audio Coding (AAC) audio"
6. ISO/IEC 13 818–7. "*MPEG-2 advanced audio coding, AAC*". International Standard
7. S. Meltzer, G. Moser, 'HE-AAC V.2 – Audio Coding for today's Digital Media Word', EBU Technical Review, January 2008
8. ITU-T Recommendation G.706 "Frame alignment and cyclic redundancy check (CRC) procedures relating to basic frame structures in Recommendation G.704".
9. J. Viterbi, 'Convolutional codes and their performance in communication systems', IEEE Trans. On Communications Technology, vol. com-19, nr. 5, October 1971
10. ETSI EN 300 797 "Digital Audio Broadcasting (DAB); Distribution interfaces; Service Transport Interface (STI), annex C
11. Arne Borsum,Encoding and multiplexing systems', DAB+ digital radio technology – Implementation and rollout, Joint WorldDAB and ASBU Webinar, Sept. 2020
12. L. P. Lundgren,Encoding and multiplexing systems', DAB+ Digital Radio Technology – Implementation and Rollout, Joint WorldDAB and ABU webinar, Kuala Lumpur, Oct. 2020
13. 'Advanced digital techniques for UHF satellite sound broadcasting', Collected papers on concepts for sound broadcasting into the 21st century, EBU, August 1988
14. B. Farhang-Boroujent,OFDM Versus Filter Bank Multicarrier', IEEE Signal Processibg Magazine, May 2011

15. Les Sabel, 'DAB System Features', DAB+ Digital Radio Technology – Implementation and Rollout, Joint WorldDAB and ABU Webinar, Kuala Lumpur, Oct. 2020
16. A. Tseng,Receiver examples – Home', DAB+ Digital Radio Technology – Implementation and Rollout, Joint WorldDAB and ABU webinar, Kuala Lumpur, Oct. 2020
17. R. Lanctot,Car Connectivity Changes Content Consumption', DAB+ Digital Radio Technology – Implementation and Rollout, Joint WorldDAB and ABU webinar, Kuala Lumpur, Oct. 2020
18. P. Rajalingham,Receivers – home', DAB+ digital radio technology – Implementation and rollout, Joint WorldDAB and ASBU Webinar, Sept. 2020

Chapter 3
Impact of the Propagation Channel on Reception Quality of the DAB Signal

Because of the large variability, unpredictability, and amount of a number of factors, the gain of the high-frequency (HF) signal path at a specific point in space is a statistical variable with probability density well described by a log-normal distribution [1]. On this basis, the minimum values of the useful signal can be evaluated. Also, on this basis, the power level of overlapping signals from interfering stations can be calculated alike the distributions of the maximum level of interfering signals. These predictions are performed for indicated points within planned coverage areas for an assumed percentage of time.

Basic values considered in the planning of ranges or coverage of signal of single transmitter in a given channel are:

- The minimum level of the useful signal

The value of this parameter is determined by the signal level that ensures high-quality reception (also in buildings) by receivers of different types and from different manufacturers in the absence of interfering signals.

- The maximum level of the resultant interference signal in points of signal estimation
- Protection of interfering transmitter to DAB

Interference caused by the interfering signal depends not only on its gain but its nature, modulation, and other parameters. Hence, the required level of the useful signal ensuring good reception in the presence of interference signal depends on the type of interfering signals.

- Free range of DAB transmitter

This is the range of the useful signal in the absence of interference for the specified transmitter.

- The interference range of the transmitter

The range of useful signal in the presence of interfering signal

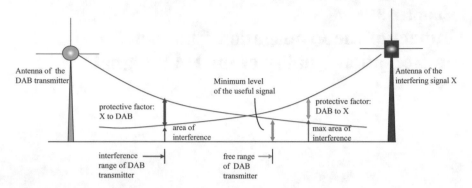

Fig. 3.1 Relations between HF signal levels considered when planning the parameters of the transmitter

These concepts are illustrated in the Fig. 3.1.

Planning coverage areas in a specified percentage of time and a percentage of locations can require the basic tools:

- Frequency allocation plan
- Signal propagation models
- Parameters of transmitter: the location of transmitter antenna and their parameters
- Maps of the area considering the topography (and morphology) of the coverage area

Coverage planning is provided for HF signals; see [2–7].

The appropriate level of the HF signal is a necessary condition of reception.

Finally, the reception quality is determined by the relationship of the OFDM subcarrier levels to interference and noise in the baseband channel. The level of the DAB usable signal in the baseband channel, especially multipath channel, is best exposed in signal phasor representation [8].

3.1   The Baseband Multipath Channel: Phasor Representation of the OFDM Signal

The task of each of the telecommunication systems is to transfer information. Information carried in the OFDM signal is contained in the phases and amplitudes of the individual subcarriers of the OFDM symbols.

Parameters of an OFDM subcarrier, amplitude and phase, can be presented in the form of a phasor of length equal to the amplitude of the subcarrier and the angle equal to its initial phase.

For the two overlapping sinusoidal signals of equal frequency, the phasor representation of the resulting signal appears as the vector sum of the component phasors. Amplitude C and phase γ of resultant phasor correspond to the subcarrier parameters

$$A \cdot \sin (\alpha + \omega t) + B \cdot \sin (\beta + \omega t) = C \cdot \sin (\gamma + \omega t)$$

where

$$C^2 = A^2 + B^2 + 2AB \cos (\alpha - \beta); \; \mathrm{tg}\, \gamma = (A\sin\alpha + B\sin\beta)/(A\cos\alpha + B\cos\beta)$$

Such phasor summation can be extended to any countable sum of sinusoidal signals with equal pulsations.

The state of each OFDM symbol of the broadband signal in a baseband channel is thus represented by set the of phasors (phasor representation), each describing the amplitude and phase of one subcarrier.

In case of the unmodulated signal, the phasor representation of the received signal reflects the current state of the propagation channel. In the case of the first path, the phasors can be treated as reference for the further paths, and all subcarrier phasors are presented as parallel. Any other path, delayed by τ relatively to the first path, is characterized by a helix line with a pitch of $1/\tau$ on a cylinder of radius equal amplitude of delayed path, as shown in Fig. 3.2 [see explanation in Annex C]. In the frequency block B, the number of revolutions of the helix line of path delayed by τ is

$$n_{obr} = \tau \cdot B,$$

or in parameters of the OFDM signal: $n_{obr} = (\tau/Tu)\cdot N$.

The path delayed by τ introduces n_{obr} local maximas to signal frequency envelope.

With the increase of τ, the number n of maximas also increases – the channel selectivity is defined by the paths with greater time delays. If the selectivity of the entire channel is not to influence the individual subchannels, the number of turns of the last significant path n_{max} must be smaller than at least half the number of subchannels:

$$n_{obr} < N/2, \; \text{or} \; T_g < T_U/2.$$

In the DAB system, the length of the guard interval T_g (maximum dispersion of time delay) is assumed as $Tg = T_U/4$.

Changing the phase of cylinder within a full revolution changes the position of the maximum in the envelope by

$$B/n_{obr} = 1/\tau.$$

so the more delayed paths have maxima within a narrower range.

Fig. 3.2 Phasor
representation of an
unmodulated OFDM
subcarriers and the helix line
representing the path of the
relative delay τ

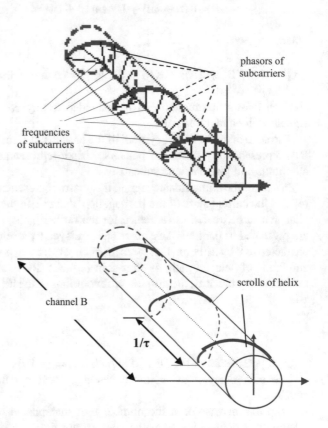

3.1.1   Relations Between the Multipath Profiles, Phasor Representation, and Frequency Envelope of Resulting OFDM Signal

Current state of propagation channel is described by the multipath delay profile (MDP), specifying the amplitudes of received paths as function of its time delays.

Each path "i" delayed by τ_i can be described in the baseband channel, in the phasor space, by the helix line with a pitch of $1/\tau_i$ and radius A_i circled by the ends of the symbol phasors of the path (see Fig. 3.3).

Relation between MDP profile and the phasor representation, sketched in the Fig. 3.3, can be summarized as:

State of the OFDM symbol = state of phasors representing information in the symbol subcarriers.

Resulting phasor of the subcarrier is the vector sum of phasors of this subcarrier from all the paths.

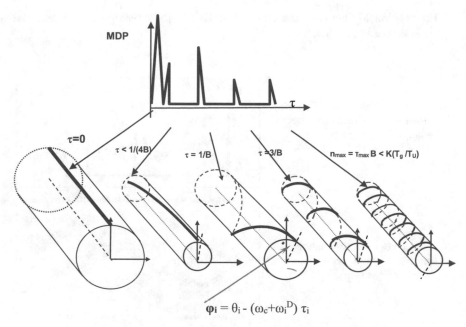

Fig. 3.3 The phasor representation of the signal paths in the baseband multipath channel

Depending on the interval to which falls the delay time of the path it contributes within the frequency block B to the formation of local minima:

- $0 < \tau < 1/4B$, the lack of a clear minimum in the amplitudes of subcarriers.
- $1/4B < \tau < 1.5/B$, one minimum in frequency block B.
- ....
- $(2n-1)/2B < \tau < (2n+1)/2B$, n minima in the block B.

Depending on the relations between amplitudes of the paths in the MDP profile, the models of propagation channels are exposed:

- *The Gauss channel*, when only one signal path is received in company of the white noise with the Gauss distribution
- *The Rice channel*, when one of the paths clearly dominates the amplitudes of the others, so the distribution of resulting signal focuses around the end of phasor of the dominant path
- *The Rayleigh channel*, when no dominant path can be distinguished and the ends of the paths phasors are distributed according to the complex Gauss distribution around the origin

The relationship between the channel MDP and the envelope of the received signal is shown in Fig. 3.4.

A. An example of Multipath Delay Profile (MDP): paths amplitudes, relative delays, and phases

B. The phasor representations of individual paths with random phases of cylinders

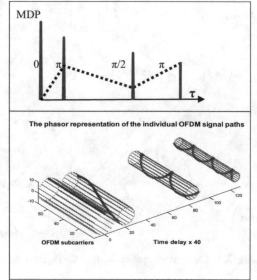

C. The phasor representation of resulting signal: the vector sum of phasors of individual paths (without a fourth path for clarity)

D. Envelope of the phasor signal: amplitudes of the resulting subcarriers – state of information data

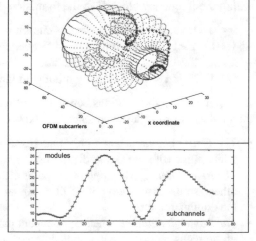

Fig. 3.4 Multipath profile→phasor representation of individual paths→resulting in phasor profile→frequency signal envelope

3.1.2   The Impact of Multipath on the Signal Reception: Examples

The phasor representation of the OFDM signal can directly facilitate the identification of sources of signal reflections that cause its degradation. It can be shown on examples of results of measurements made in and around Warsaw. The signal envelope was measured for the radio DAB signal broadcasted from center of the city.

Below each of presented examples consists of three graphs:

- On the upper chart, we have the envelope of DAB radio signal between the FM signals of analog stations.
- In the middle chart, a multipath delay profile is presented: a direct path and the delayed path with the parameters chosen so that the shape of the resultant signal is like the DAB signal.
- The bottom signal shows the frequency envelope of the resultant signal (Fig. 3.5).

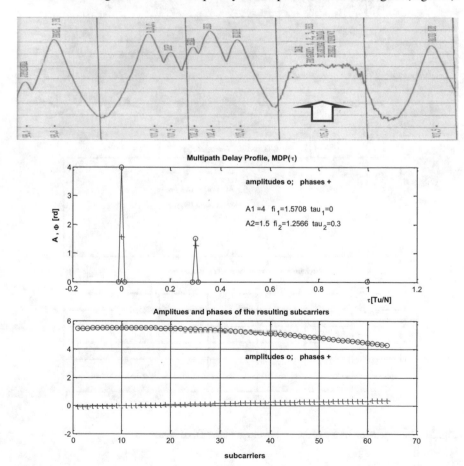

Fig. 3.5   Interpretation of the shape of the envelope of the DAB signal spectrum between UKF FM signals

Another example of the DAB signal is shown in the Fig. 3.6 (long vertical lines result the measurement method: sweeping the guard interval of the OFDM frames). The course of signal envelope can be modelled by the multipath profile of the central plot. The sum of phasor charts of both paths on the lower graph corresponds to the measured signal.

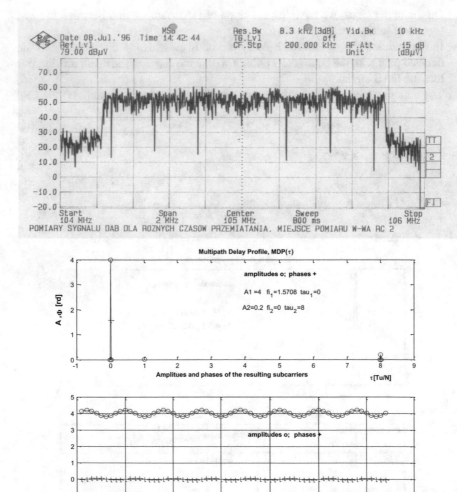

Fig. 3.6 Interpretation of the spectrum envelope of the DAB signal

Degradation of low-frequencies in the DAB signal on top of Fig. 3.7 is the result of summation of the direct signal (LOS) and reflected path with the parameters shown in the central plot. The effect of aggregation of phasor charts of both paths is shown in the bottom graph.

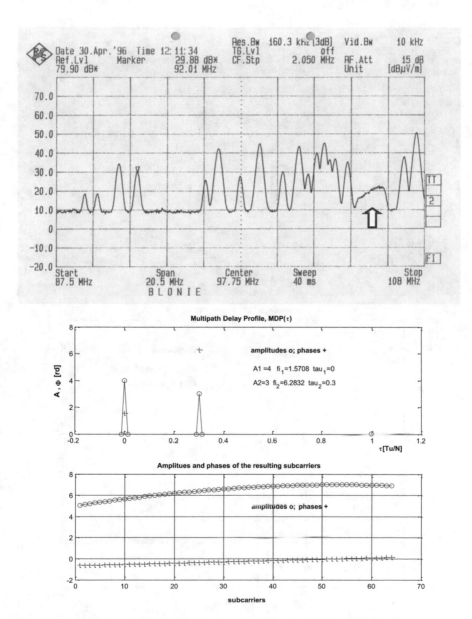

Fig. 3.7 Interpretation of the DAB signal spectrum with suppressed low frequencies

The degradation of both the low and high frequencies in the envelope of the DAB signal is presented in Fig. 3.8. Interpretation of the shape of the spectrum as a result of the sum of phasor models of direct LOS signal and a reflected path with the parameters of the MDP profile (middle graph) explains the course of the signal spectrum on lower graph.

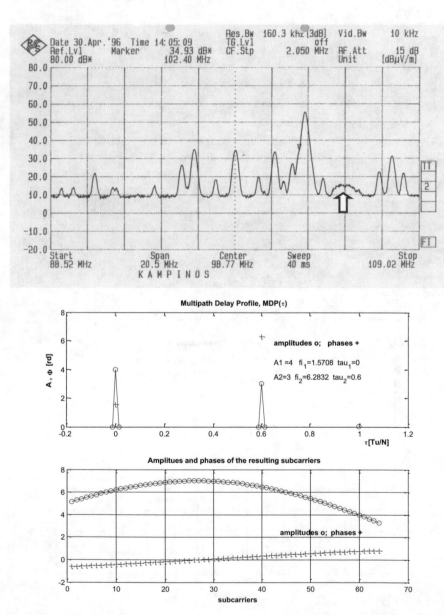

Fig. 3.8 Interpretation of the shape of the spectral envelope of the DAB signal from the region of Warsaw

In the case of two or three OFDM signal paths, a simple shape matching of the resultant signal spectrum obtained from measurements allows to determine parameters of these paths: amplitudes, phases, and time delays. Knowledge of the time delay of each path allows for the appointment of an ellipse which is the locus of the tracks of the specified delay. Using the map (digital) of the analyzed coverage area with indicated sizes of buildings (number of floors) allows thus to determine the source of reflections affecting the poor reception.

These data allow for determination of the places of reflections and thus the orientation of the receiving antenna for limiting the impact of degrading reflections on stationary reception (Fig. 3.9).

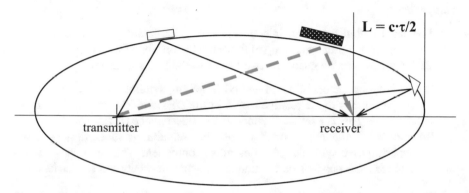

Fig. 3.9 The ellipse as the locus of points of signal reflections with time delay τ relative to the LOS signal

3.2   Inter-symbol and Inter-tone Interferences

Selection of the OFDM signal parameters is intended to avoid inter-symbol interference resulting from the occurrence of signal paths outside the guard interval. The length of the guard interval of OFDM symbols should be chosen to accommodate dispersion of the time delays in the area. However, since the introduction of the guard interval reduces the throughput of the system, it is not justified to extend its length for maximal delayed paths, appearing only in exceptional conditions.

The phasor representation of the OFDM symbol is described below in such cases.

3.2.1   Phasor Representation of the OFDM Paths Outside the Guard Interval

The appearance of the OFDM signal paths outside the guard interval causes:

- Disruption of continuity of each subcarrier sinusoid in the FFT processing period: the modulating factors of the OFDM symbol are summed up with factors of the next symbol.
- Disruption of orthogonality of subcarriers, so demodulating one subcarrier simultaneously introduces contributions from other subcarriers.

As a result, the demodulated coefficient of the k-th subcarrier is made up of components:

(a) The paths within the guard interval are giving modulating factors of the k-th subcarrier in the n-th symbol $q_k(n)$ (with the distortions of the channel).
(b) Each path outside the guard interval brings to the k-th subcarrier:

  (a) Modulating factors of a current and following symbol $\{q_k(n)\}$, $\{q_k(n + 1)\}$ proportionally to the relative length of the DFT window occupied by those symbols. This is a result of inter-symbol interference (ISI).
  (b) Sum of modulating symbols of the remaining subcarriers – from superimposed symbols – forming noise component. This inter-carrier interference (ICI) depends on the frequency separation between subcarriers.

Graphically, this situation is represented in the Fig. 3.10 and templral relations in Fig. 3.11.

In Fig. 3.10c, the corresponding graphs are shown for the example from Fig. 3.10a with delay profile of Fig. 3.10b. The component of the noise created by the ICI component is here missing.

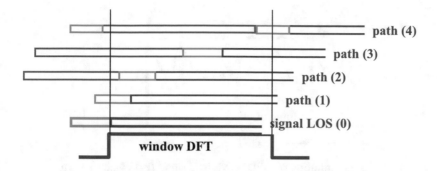

A. Positions of OFDM symbols of delayed paths in the DFT window

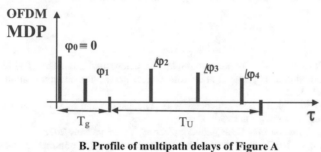

B. Profile of multipath delays of Figure A

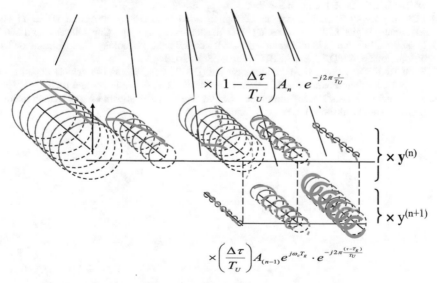

C. Phasor representation of the OFDM symbol in the case of paths {0,1} within, and {2,3,4} out of guard interval

Fig. 3.10   Phasor representation of the signal profile MDP of paths also outside the guard interval

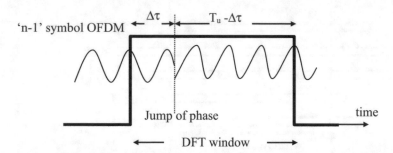

The contribution of OFDM symbols n and (n-1) when $\tau > Tg$
in the DFT transform window

Fig. 3.11 Temporal relations of path with $\tau > Tg$ within the DFT window

References

1. T. O'Leary, J. Rutkowski, 'Combining multiple interfering field strengths: The simplified multiplication method and its physical and mathematical basis', Telecommunication Journal, vol. 49, November 1982
2. ETSI TR 101 496-3 "Digital Audio Broadcasting (DAB); Guides and rules for implementation and operation; Part 3: Broadcast network",
3. EBU, Tech 3391 "Guidelines for DAB network planning", Geneva May 2018
4. R. Brugger, K. Mayer,'RRC-06 – Technical basis and planning configurations for T-DAB and DVB-T', EBU-UER Technical Review, 2005, pp. 66-73
5. ECC Report 49,'Technical criteria of digital video broadcasting terrestrial (DVB-T) and terrestrial – Digital audio broadcasting (T-DAB) allotment planning', Copenhagen, April 2004
6. H. Zensen,'Real world implementations', DAB+ digital radio technology – Implementation and rollout, Joint WorldDAB and ASBU Webinar, Sept. 2020
7. ETSI TS 103 461 "Digital Audio Broadcasting (DAB); Domestic and in-vehicle digital radio receivers; Minimum requirements and Test specifications for technologies and products"
8. M. Oziewicz, 'Phasor Description of the COFDM Signal in a Multipath Channel; Sixth International OFDM-Workshop (InOWo'01), Hamburg, 2001

Chapter 4
Single-Frequency Networks (SFN)

The ability to create a network of transmitters operating on the same carrier fre-quency, i.e., the single-frequency network, is closely related to the OFDM channel modulation system developed for DAB digital radio.

4.1 Concept of the Single-Frequency Network (SFN)

The SFN or the single-frequency network is a network of time-synchronized trans-mitters broadcasting the same signal at the same frequency block and on the same carrier frequency.

The concept of SFN stems from the properties of the OFDM encoder. As highlighted in Sect. 2.8, the parameters of OFDM encoder are chosen so that the DAB system is immune to the effects of multipath propagation. If the receiver of DAB system can decode the OFDM signal in the case of overlapping signals due to multipath propagation, it can therefore be concluded that the reception of overlapping signals from different DAB transmitters also will run, provided that:

– The transmitted signal (the program) will be identical in each transmitter
– Time delays of signals from different transmitters will not exceed the length of the guard interval between OFDM frames

The conceptual difference between multiple-frequency networks (MFN) and single-frequency network (SFN) is illustrated in Fig. 4.1, where the signal in SFN is received in the baseband channel as the sum of the paths with different mutual delays.

Depending on the area of SFN service coverage, the following are distinguished:

– Local network SFN
– SFN regional network
– National network SFN

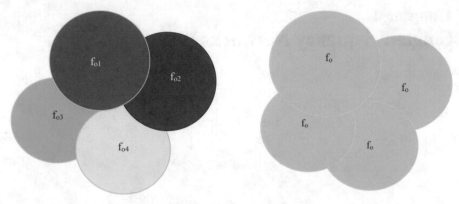

**A. Comparison of signal coverage of Multi Frequency Network
and Single Frequency Network**

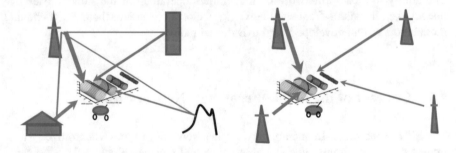

▪ **The baseband signal received in a multipath channel and a 4 SFN network**

Fig. 4.1 The concept of MFN network and 4 SFN DAB network

Advantages of the single-frequency networks:

- Frequency planning to cover a specific area with program is considerably optimized. For example, a broadcasting program covering territory of Poland in the FM system requires a block of frequencies around 4 MHz for the deployment of not interfering transmitters. In the same frequency block in the DAB+ system, two blocks can be transmitted, each with about 20 different programs with national coverage.
- Reception of the selected program in a car along a route in a wider area without retuning the receiver between the different transmitters.
- Reception suppressed by obstacles from the direction of one transmitter is compensated by signals of other transmitters in the network.

Planning the SFN network requires – besides estimation of the signal levels in points within coverage area – coordination of the network topology (the distances between the antennas of transmitters) with a guard interval of the OFDM modulator, in order to control the delays of the signals from the neighboring transmitters.

Signals from more distant transmitters will be outside the guard interval range creating weak interference. The correlation between the length of the guard interval limiting the throughput of the signal, the level of ISI and ICI interference, transmitters radiating powers, and topology of transmitters network is a new problem with planning SFN compared to planning multiple-frequency network (MFN).

4.2 Rules for Planning the Single-Frequency Network

Technical rules and examples of planning SFN networks are presented in documents [1–4] and webinar presentations [5–10].

The Doppler frequency shift f^D of DAB receivers for allowed limit of vehicle speed v_{max} depends on the carrier frequency of network F_c:

$$f^D = F_c \cdot v/c$$

If these shifts are not to significantly affect the quality of reception, they should not exceed 5% of intercarrier interval Δf. The orthogonality relation of the OFDM subcarriers requires the relation $\Delta f = 1/T_U$, where T_U is the length of the OFDM useful field. So, greater intercarrier distances mean shorter useful fields.

In consequence, also the guard intervals should be shorter, because the length of guard interval should not exceed $\frac{1}{4} T_U$. This is due to the requirement of flat characteristics of the OFDM subchannels. For this aim, the number of subcarriers N should be at least four times greater than the number n of turns of the helix line describing the farthest meaning path within the guard interval:

$$T_U/4 \geq T_g$$

The DAB blocks from the higher-frequency bands require symbols with shorter guard intervals. In the case of SFN networks, this means shorter intervals between antenna masts and therefore denser network of transmitters to provide signal coverage in a specified area.

On the other hand, the denser network does not require high-power transmitters, as opposed to the network of large mutual distances.

Indicated relationships are depicted in Fig. 4.2.

In the case of the DAB+ system, the range of carrier frequencies (VHF band III) and OFDM parameters (mode I) is fixed. Thus, SFN planning includes configuration and power of antennas and network synchronization.

Internationally agreed frequency blocks in each country are then planned in detail based on the real location of the planned broadcasting stations and their parameters:

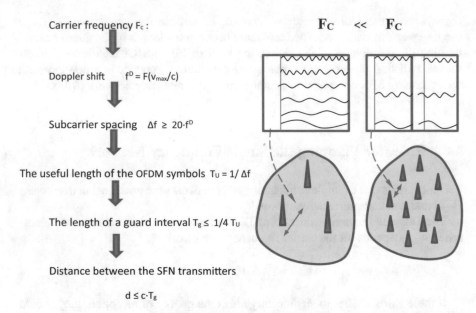

Carrier frequency F_c :

Doppler shift $f^D = F(v_{max}/c)$

Subcarrier spacing $\Delta f \geq 20 \cdot f^D$

The useful length of the OFDM symbols $T_U = 1/\Delta f$

The length of a guard interval $T_g \leq 1/4\, T_U$

Distance between the SFN transmitters

$d \leq c \cdot T_g$

$F_C \quad << \quad F_C$

Fig. 4.2 Selection of the OFDM encoder mode in mobile reception in SFN networks. Relations: frequency range > mode of OFDM > distance between antennas of SFN

– Geographic coordinates
– The relative heights of antenna masts
– The powers of the transmitters
– Characteristics of the antennas

In the same way, there is organized a plan of the designed network, whose various elements are then consulted and negotiated with national and foreign administrations of the neighboring blocks. The parameters that should be analyzed in the design of SFN networks are the levels of field strength on the border of the network. According to the remark in Sect. 1.1, a digital signal with a gain below the required minimum level is rapidly degraded. Therefore, for the DAB network planning, it is assumed that the signal must cover 90% of the specified area (for analogue broadcasting, 50%) within a 50% of transmission time (as for FM).

As a result of mutual agreement, there is formed the final structure of the planned single frequency network.

The range of a single frequency block is characterized by the following parameters:

– The required minimum intensity of EM fields
– The maximum permissible intensity of interfering fields:

• DAB by DAB (between blocks of different programs)
• TV, FM, ... by DAB
• DAB by TV, FM, ...

The rate of EM field decreasing outside the SFN is determined, alike for FM, by the propagation curves of ITU-R Recommendation 370 with the respective corrections.

The block with the same carrier frequency can be used in another SFN network, distant far enough from the first one, to avoid mutual interference. It is assumed that this distance ("coordination distance") is about 80 km in the VHF band.

4.2.1   The SFN Synchronization

The concept of SFN of transmitters is conditioned by the time synchronization of identical signal transmission in each of the network transmitters. If the signals from different SFN transmitters are not to generate interferences in the receiver, their mutual delays should be generally within the guard interval.

The DAB signal sent to the transmitters of SFN network requires mutual synchronization due to the following:

(a) The times of signal reception in differently located transmitters are different (Network Padding Delay).
(b) The time of signal processing at the transmitters from different manufacturers may vary (Viterbi coding, OFDM modulation channel, output HF circuitry).
(c) The time of signal transmission in fider between the transmitter and the antenna depends on the fider length.

To correct the differences in the time of broadcasting, the DAB signal in different SFN transmitters is corrected using the timestamps in the ETI signal supplying the SFN (see Sect. 9.1). The timestamps allow to compensate time differences in the network of up to hundreds of milliseconds.

4.2.2   Identification of Transmitters within SFN: Codes TII

A signal from the DAB server is supplied in the single-frequency network to the individual transmitters. After demodulation, the signal at each point within the network coverage has the same content in the DAB identical frames. The signal transmitted in this way makes it impossible to distinguish in the receiver within which transmitter range it is located. But a class of information, like advertisements or emergency information, is naturally associated with the location limited to the area of indicated transmitter, or even smaller area defined by the geographic coordinates. In case of traffic messages, or notices related with selected locality, the information should be decoded only by receivers from the indicated area. For this purpose, the individual identification codes TII (Transmitter Identification Information) can optionally be allocated to individual transmitters within the SFN.

The TII codes consist of two numbers (p, c), of which the main identifier p (Main Id) in the range 0–69 indicates the main transmitter of subnet to which the transmitter belongs and the secondary identifier c (Sub), in the range 1–23, determines the relative number of the transmitter in the subnet identified by the main transmitter. Indexes (p, c) identifying the transmitter are encoded in the system of subcarriers of the zero OFDM symbol in synchronization channel at the start of each DAB frame [1]. The subcarriers representing codes are chosen in such a way that signals of subnet transmitters do not introduce mutual interference.

If the transmission of a file or message is associated with a specific geographic location, or specified location of the transmitter, then the DAB receiver within SFN network area compares the code of transmitter with the code in the message. If the codes agree, the message is decoded.

The decision on the introduction of transmitter identification codes TII is taken by the operator of DAB multiplex and sent to the transmitter network in the transport interface frame ETI (Ensemble Transport Interface) in the phase field of the frame (see Sect. 9.1).

4.2.3 Role of Gap Fillers

Within the SFN coverage area, there can appear places, where due to the low ground, or shadows caused by the terrain obstacles or buildings, the gain of received signal is not sufficient for reception of radio program within a given percentage of the time. It can happen that the correct reception is not possible at all. In such places, there can be applied stations reinforcing SFN signal in the same frequency block. Such complementary stations, called gap fillers, receive the transmitted weak SFN signal and, after amplification, retransmit it in the assumed area using antennas with adequate angle parameters. Besides the level of the emitted secondary signal gain, the time delay between the two signals should also be considered when designing parameters of the station. In real conditions, these relations will depend on topography of surrounding terrain and obstacles which obstructs the signals. However, in order to illustrate the problem, a simple model in Fig. 4.3 is sufficient. Here δt is the time of signal processing in the gap-filler station, Δ_t the relative delay between the two signals, R the radius of the coverage area of the station, and c the speed of the signal.

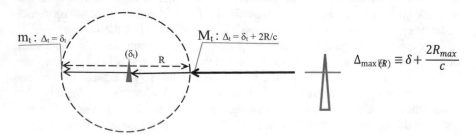

Fig. 4.3 Temporal relations of signals in the area of complementary station

In the coverage area common for both signals (the primary one from transmitter and secondary from gap-filler station), the primary signal will be received as the first, because the signal from gap-filler station will be delayed at least by the time δt of the signal amplification. In addition to this time, delay should be added resulting from the difference of signal paths to the reception point. On the above sketch, there are marked point with zero path difference (m_t) and a point (M_t) with maximum difference of signal paths. To avoid the inter-symbol interference, the time delay between signals should not exceed a length of the guard interval T_g of the OFDM symbols, so in the point M_t

$$\Delta_{max\,(R)} \equiv \delta_t + \frac{2R_{max}}{c} < T_g, \text{from where } R_{max} < \frac{c \cdot (T_g - \delta_t)}{2}$$

To avoid the inter-symbol interference in the area around point M_t, the signal emitted from the gap-filler station should be under the gain of primary signal for the value of protection factor of DAB to noise.

4.2.4 Local Programs in the SFN Networks

It can happen that within the SFN, some programs are designed only for specified local regions. So, at least during appointed time, such programs should be emitted only in these regions. Realization of such emissions can be organized under certain conditions.

If one of the subchannels in the signal broadcasted in the single-frequency network will be muted, in this time, in the local networks of non-intersecting coverages can be allowed broadcasting of its own independent programs. This situation is illustrated in Fig. 4.4. Transmitters A, B, and C form a single-frequency network broadcasting synchronized program containing subchannel of silence. In this subchannel, the emitters A and C can broadcast its own independent local programs. Transmitter B is not broadcasting any program in this subchannel. Another method was proposed by Rai Way from the Rai Research Centre, see [1, Annex k].

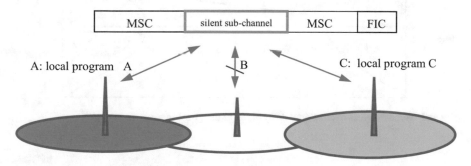

Fig. 4.4 The concept of local programs within the coverage of SFN

4.3 SFN Signals in the Baseband Channel

Predictions of the signal gain at the points of specified SFN area based on heuristic data and statistical evaluations of high-frequency signal lead to necessary but not sufficient conditions of proper reception. The reliable signal reception depends on the relations of subcarriers to noise in the baseband channel. The values of baseband signal can best be presented in the phasor representation, exposing levels of information subcarriers.

4.3.1 Intra-network Signals in the Large SFN Model Network

A. Inter-symbol Interference in 2 SFN

As pointed out in Sect. 3.2, any path of time delay exceeding the guard interval contributes to the inter-symbol and intercarrier interference. In the case of a single-frequency network (SFN), the signal received by the receiver is the sum of signals from the individual transmitters with different time delays. In the 2 SFN with transmitters at a distance X, the interference will appear within hyperbola (Fig. 4.5), *the apex* of which is defined by the following coordinates:

$$r_1 + r_2 = X$$
$$r_1 - r_2 = c \cdot T_g,$$

so

$$r_2 = X/2 - c \cdot T_g/2$$

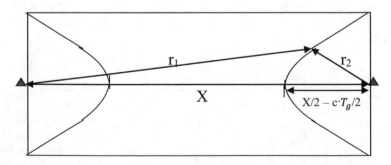

Fig. 4.5 Hyperboles of relative time delay equal to the guard interval T_g in 2 SFN

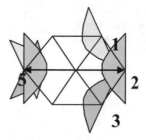

Ranges of interference around the
transmitters caused by signals
delayed over guard interval of
delayed transmitters

Fig. 4.6   Sketch of 7 SFN with areas of internal interference ISI/ICI

Hence it follows that neither in the network 2 SFN with $d \equiv X = c \cdot T_g$ nor
equilateral flat SFN network of 3 antennas at distances $c \cdot T_g = d = X$, the internetwork
interference does not occur. Only in the larger networks there appear signals from
distant transmitters $(X > cT_g)$ with time delays outside the guard interval.

Figure 4.6 shows a model of 7 SFN.

The shaded areas indicate the presence of interference ISI/ICI.

B. Examples of Baseband Signal in Model 6 SFN: Computer Simulations

Below are presented examples of envelopes of resulting signals in points of
model flat 6 SFN (see Fig. 4.7). These examples indicate dependence of the signal
envelope on the time delays of signal paths [11].

The assumed parameters of model network: all the masts height $H = 150$ m;
radiation power of every transmitter 10 kW; carrier frequency Fc = 200 MHz; distance
between transmitters d = 10 km. Estimation points (1)–(8) are evenly distributed in
points on the arrow every 1200 m starting outside the near field of the transmitter A.

The estimations were provided for signal DAB with $N = 192$ subcarriers,
$T_U = 125$ us, and $T_g = 32$ us. An intensity of the direct and reflected HF signal
from any transmitter at distance r_n at a height $h = 2$ m over earth was assumed as

$$ E(r_n) = \frac{E(P,H)}{r_n^2}, \text{where } E(P,H) = 2.18 \cdot sqrt(P) \cdot H \cdot h \cdot Fc/300 $$

Delay times of signals from different transmitters were ranked with increasing
length allowing to separate the useful signals and the interfering parts of ISI and ICI
for delays more than a guard interval. Delay times here are calculated after
subtracting the delay of the first path giving the relative delays.

Knowing the signal bandwidth and the relative time delay of path, the number of
maxima of the envelope of that path [Annex C] is determined as

$$ n = \tau \cdot B = \tau \cdot N/Tu = 1.56 \cdot \tau \text{ [us]} $$

Phases of individual paths (cylinders) were chosen statistically between 0 and 2π.
Also modulating factors of the following symbol – necessary for calculating ISI and
ICI – were chosen statistically. Resultant envelopes of signals depend on amplitudes
and phases of the individual paths as was discussed earlier (see Fig. 3.4).

Fig. 4.7 Configuration of transmitters of the model 6 SFN. An arrow indicates the route of signal estimation points

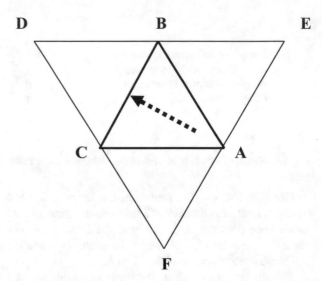

In Figs. 4.8(1), 4.8(2), 4.8(3), 4.8(4), 4.8(5), 4.8(6), 4.8(7), and 4.8(8), the waveforms of the useful ISI and ICI signals for the subcarrier frequencies in the subsequent estimation points (1)–(8) are presented. The delay times of meaning paths and adequate number of their maxima are marked below.

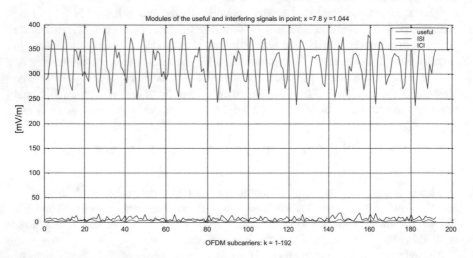

Fig. 4.8(1) Envelopes of signals of 6 SFN in point 1
Relative time delays: 0   19.3   19.29   26
Adequate path maxima: 0   30.06   30.09   40

The average level of the signal (~ 320 mV/m) is determined by the path from transmitter A, the reference path. Waving of envelope is caused by interfering delayed signals from transmitters B and C (delays and amplitudes of paths from B and C are similar to resulting interferences). The signals from other transmitters (nearest E and F) have no meaning part in the resultant signal.

The second estimation point is further away from the vertex A, so the average level of the signal envelope is smaller (~ 150 mV/m). The increased amplitudes of signals from transmitters B and C increase the interferences.

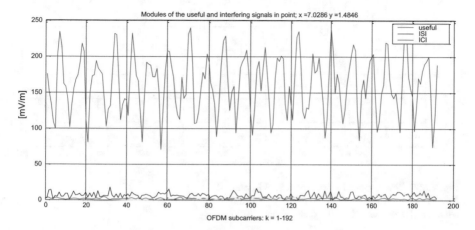

Fig. 4.8(2) Envelopes of signals of 6 SFN in point 2
Relative time delays: 0   14.05   14.05   24
Number of path maxima: 0   21.92   21.92   37

In the third estimation point, the average level of the signal envelope (~ 100 mV/m) is still dictated by the path from transmitter A. Signals from transmitters B and C cause the slow waving with 14 maxima, and the fast waving of envelope is the result of interfering signals from transmitters E and F.

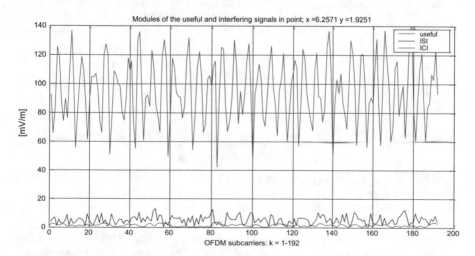

Fig. 4.8(3) Envelopes of signals of 6 SFN in point 3
Relative time delays: 0   8.96   8.98   21.6
Number of path maxima: 0   13.97   14.00   33.07

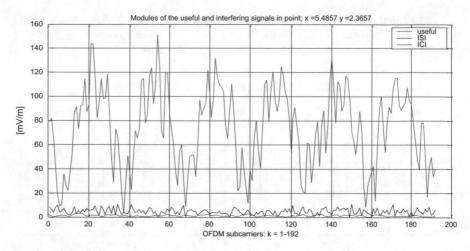

Fig. 4.8(4) Envelopes of signals of 6 SFN in point 4
Relative time delays: 0   4.09   4.14   19.85
Number of path maxima: 0   6.38   6.46   30.97

An average level of amplitude is appointed by the first path from A, transmitters B and C create the six main waves of amplitude, and the last meaning paths from E and F decide about fast changes of envelope.

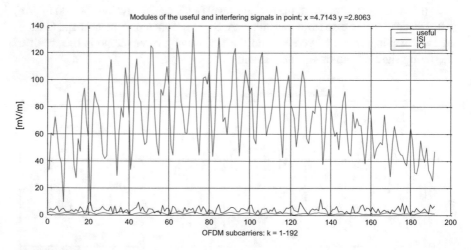

Fig. 4.8(5) Envelopes of signals of 6 SFN in point 5
Relative time delays: 0   0.07   0.49   18.35
Number of path maxima: 0   0.1   0.76   28.63

The three first paths are close to one another (see delays). This means that the estimation point is near the center of triangle ABC. The first and second paths decide about the meaning value of signal, and the third path decides about its arch (less than one turn in the phasor representation of this path). The interfering paths from transmitters D, E, and F create the fast waving of the envelope.

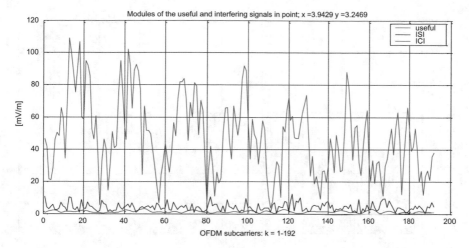

Fig. 4.8(6) Envelopes of signals of 6 SFN in point 6
Relative time delays: 0   0.10   4.72   16.65
Number of path maxima: 0   0.16   7.36   25.97

Here the first paths come from transmitters B and C, so delays between them are smallest. Signal from transmitter A decides about main maxima (7), and the path from transmitter D gives the fast changes of envelope.

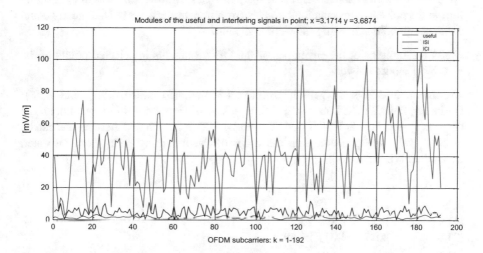

Fig. 4.8(7) Envelopes of signals of 6 SFN in point 7
Relative time delays: 0   0.14   8.49   14.51
Number of path maxima: 0   0.22   13.24   22.64

Here interpretation is alike that in Fig. 4.8(6), with its new parameters.

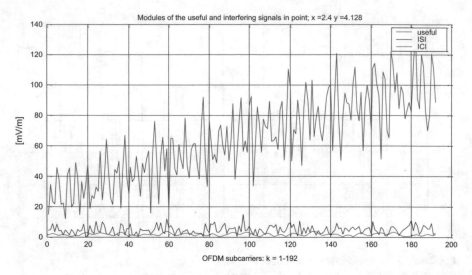

Fig. 4.8(8) Envelope signals of 6 SFN in point 8
Relative time delays: 0   0.17   11.75   11.84
Number of path maxima: 0   0.26   18.33   18.47

The estimation point 8 lies close to center of line B-C. The paths from B and C appoint the average signal value and the slope of envelope (one fourth of a single turn in the phasor representation); signals from transmitters A and D, at nearly equal distance, give fast changing interfering signals.

Fig. 4.8 Examples of the envelopes of the OFDM signals in selected points of the 6 SFN model network

Comments and Figs. 4.8(1), 4.8(2), 4.8(3), 4.8(4), 4.8(5), 4.8(6), 4.8(7), and 4.8(8):
Gathering together the four smallest time delays of signal paths in Figs. 4.8(1), 4.8(2), 4.8(3), 4.8(4), 4.8(5), 4.8(6), 4.8(7), and 4.8(8), it is seen that these results can be divided into parts depending on the order of the incoming paths from indicated transmitters (in brackets):

	(A)	(B)	(C)	(E/F)
(1) >	0	19.27	19.29	25.79
(2) >	0	14.05	14.05	23.59
(3) >	0	8.96	8.98	21.6
(4) >	0	4.09	4.14	19.85

	(A)	(B)	(C)	(D/E/F)
(5) >	0	0.07	0.49	18.35

	(B)	(C)	(A)	(D)
(6) >	0	0.10	4.72	16.65
(7) >	0	0.14	8.49	14.51
(8) >	0	0.17	11.75	11.84

In estimation points (1)–(4), the first signal comes from transmitter A – it is signal with reference time delay. Next appear signals from pair of transmitters B and C; theoretically simultaneously, but because of the limited accuracy of the distances, the delay times may also little vary. The last time delay comes from first of transmitters E and F.

In point (5), close to central of triangle ABC, the signals from transmitters A, B, and C are close, alike signals from transmitters D, E, and F.

In estimation points (6)–(8), the first signals come from the pair of transmitters B and C. The distance from transmitter A grows up to the distance from transmitter D in the last point (8).

4.4   Conclusions

A. Deterministic nature of the model shows the signal as a "snapshot" of the baseband channel at selected points. Significant impact on the resulting envelope have phases of individual paths (angles of cylinders in the phasor representation of MDP profile) taken in above examples as a set of random values.

Over time this image may be subject to some variations – relations between amplitudes of the paths in reception point undergo some shifts with the change of propagation parameters – but the course of the field will retain its character.

B. The areas at risk of lack of signal reception – despite the presence of signal paths – are places where the delays of paths with an even signal level differ close to zero. To avoid such situations, the signal coverage planning can take into account:

(a) Differentiation of the power levels of the leading signals at the delay reset points

(b) Moving the time delay reset point to the region of different signal levels by proper time delays from transmitter paths

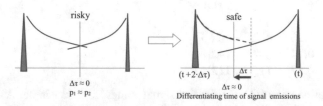

The possibility of differentiating the signal level and regulating times of signal emission in the transmitters can be taken into account when planning the SFN network in order to optimize the signal coverage.

References

1. EBU, Tech 3391 "Guidelines for DAB network planning", Geneva May 2018
2. EBU BPN 003,'Technical bases for T-DAB services network planning and compatibility', 2013
3. 'Digital Audio Broadcasting Eureka-147. Minimum Requirements for Terrestrial DAB Transmitters', Prepared by WorldDAB, September 2001
4. Eureka Project 147,'Digital Audio Broadcasting System. Guidelines for Implementation and Operation', vol. II: System features – Implementation and Operation, prepared by the Joint Eureka -147 DAB WG1/EBU Task Force on System Standardization,
5. J. Devens, 'Planning of Single Frequency Networks', ITU/EBU Workshop on Digital Broadcasting, Sofia, June 2004
6. Les Sabel,'SFN design and examples', Joint WorldDAB and ASBU DAB+ technical webinar, Sept. 2020
7. J. Hirigoyen,'DAB network design', DAB+ digital radio technology – Implementation and rollout, Joint WorldDAB and ASBU Webinar, Sept. 2020
8. Les Sabel,'Transmission system overview', DAB+ Digital Radio Technology – Implementation and Rollout, Joint WorldDAB and ABU webinar, Kuala Lumpur, Oct. 2020
9. R. Redmond, 'Modern Transmitters', DAB+ Digital Radio Technology – Implementation and Rollout, Joint WorldDAB and ABU webinar, Kuala Lumpur, Oct. 2020
10. Les Sabel,'Coverage and interference planning', DAB+ Digital Radio Technology – Implementation and Rollout, Joint WorldDAB and ABU webinar, Kuala Lumpur, Oct. 2020
11. M. Oziewicz,'Phasor description of the OFDM signal in the SFN network', IEEE Trans. Broadcast., vol. 50, no 1, pp. 63-70, March 2004

Chapter 5
DAB Multiplexes

5.1 Concept of Multiplexes

The concept of the multiplex includes:

1. The cumulative network of programs and services that use a common frequency block.
2. The common frequency block modulated as the OFDM signal, in the Specifications referred to as the ensemble signal. This signal contains multiplexed programs and services organized in subchannels – so the term used simultaneously is the multiplex signal.
3. The common coverage area of multiplex signal of transmitters operating in the Single Frequency Network (SFN).

A sketch of a centralized organization of multiplex with its distinguished elements is shown in Fig. 5.1.

Signal of multiplex is created by programs and services that use a common block of frequencies and the joint broadcasting network. The decisive element in the organization of multiplex signal is the multiplexer (Sect. 2.8). A multiplexer is a device linking signals of programs and services of different operators, with different bit rates, in the main signal in the transmission channel, i.e., the Main Service Channel. The aggregated output signal of multiplexer is fed to transmitters of the single-frequency network, one or several SFNs. In each transmitter, the logical signal of multiplex undergoes the OFDM modulation and, after synchronization with an entire network, is jointly broadcast [1].

The individual subnets of multiplex can be organized in a flexible manner.

Ad. 1. The bandwidth of the DAB+ frequency block (1.536 MHz) requires inclusion of several parallel programs at the same time. Delivery of individual programs and services to the DAB server may be realized through different telecom links. Description of the service transmission to DAB server is presented in Chap. 8.

M. Oziewicz, *Digital Radio DAB+*, https://doi.org/10.1007/978-3-030-66478-7_5

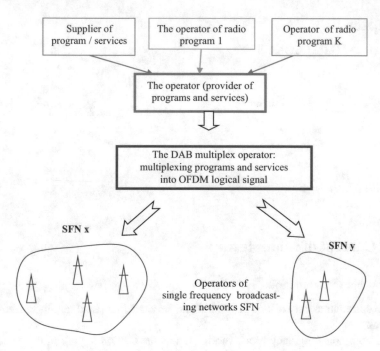

Fig. 5.1 Scheme of transport of multiplex programs and services from the supplier to the transmitters

Ensemble operator collects programs of individual operators along with the parameters of transmission closer discussed in Chaps. 6 and 7.

Ad. 2. The multiplex organization, aside from the centralized form shown in Fig. 5.1, may be configured in the dispersed form, with the intermediate multi-plexes connecting portion of services or programs [2].

Ad. 3. To bring the multiplex signal to the transmitters of each single-frequency network (SFN), different telecommunication links can be used. Transport orga-nization of multiplex signal is described in Sect. 9. Signal processing in the convolutional encoder and the channel modulator OFDM takes place in individ-ual transmitters. The aim of such solution is to avoid transmission of the incremented signal rate of the redundant code of convolutional encoder.

Planning of multiplex is connected with:

– Internationally agreed blocks of the DAB frequencies
– The authorized coverage areas of SFN networks following the cross-border arrangements
– Compatibility with existing broadcasting services
– Adjacent channel issues
– Inter-block interference
– Re-use distance for T-DAB allotment plan [3]

5.2 Planning the Coverage Area for Multiplexes

Unlike the channels in the analog broadcasting, the fundamental frequency units in system DAB are the frequency blocks of 1.5 MHz. Designing the digital radio in Europe started with the adaptation of a grid of 4 frequency blocks as 4 quarters of a single television frequency channel [4].

In its initial proposal, any multiplex can be assigned a coverage area from the proposed hypothetical grid. The four frequency blocks of multiplexes can be arranged in a checkerboard pattern of non-overlapping areas named A, B, C, and D, as in Fig. 5.2. Such division can gradually cover any area.

Allocation of additional TV channels for digital radio offers the possibility of further developing DAB multiplex networks. Based on this assumption, the initial allocation of frequency blocks for each country is adapted to the real conditions considering:

- Limitations resulting from the international agreements
- Limitations resulting in the bilateral cross-border arrangements between neighboring countries
- Compatibility with other telecommunication systems within a country

The arrangements of the ITU Conference in Wiesbaden (1995) revised in Maastricht (2002) and Constanta (2007) granted to system DAB the frequency bands in VHF: 174–230 MHz and 230–240 MHz [5].

The frequency range of transmitters for terrestrial broadcasts is located in a technically reasonable limit

$$174 - 240 \text{ MHz (VHF range)}$$

tunable every 16 kHz.

Unlike in the analog radio, where the basic concept is the frequency channel corresponding to one program, in the DAB digital system, the basic units are the blocks of frequency size 1.536 MHz. The gross bit throughput of each block is 2304 kbit/s.

Fig. 5.2 The initial proposal of multiplexes A, B, C, and D assignment to the geographical regions

A	C	A	C	A
B	D	B	D	B
A	C	A	C	A
B	D	B	D	B
A	C	A	C	A

Fig. 5.3 An example of an international plan of allocation blocks in L-band (DMB "Eureka," in February 2013, according to working Report ECC 188)

In one frequency block simultaneously can be emitted about ten programs in the DAB, twice more programs in the DAB+ system, and additionally some added information.

The number of programs transmitted in the block may be different. It depends on the compression ratio of sound channels and capacity of the value-added services. Certain types of digital information at the receiver can be presented as a correlated object. Such correlated components together form a multimedia object. The individual components are called monomedia.

The main carrier of programs and information in the stream of DAB logical frames is the logical main service channel (MSC). The information on division and organization of MSC is included in the logical Fast Information Channel (FIC), decoded at the receiver about 384 milliseconds earlier than the main service. This is due to omission of the time interleaver of logical frames in the FIC channel, which delays the transmission process. Adoption of constant efficiency 1/3 for convolutional encoder in the FIC channel secures robust transmission. A faster transmission in the FIC channel is required in order to prepare the demultiplexer

Table 5.1 The DAB+ signal mode I parameters [1]

Parameter	Value
Sampling frequency	48 kHz
Frame TF	96 ms
The number of OFDM symbols in frame	76
Null symbol T_0	1.297 ms
Useful (orthogonal) field t_{est}	1 ms
Guard interval D	0.246 ms
Full symbol T_{so}	1.246 ms
Number of subcarriers	1536
Intercarrier spacing	1 kHz

to the deconvolution of frames of individual programs and the frames of digital information.

Basic parameters of the output DAB+ frames are presented in Table 5.1.

A single transmitter covers with the program the local area. If the program is to cover a larger area, the coverage requires a network of transmitters. The most economical solution, from point of view of the optimal application of the spectrum, is a single-frequency network (SFN) [6]. Such network requires the transmission of formatted DAB multiplex signal (DAB ensemble) to each of the transmitters in a coordinated, synchronized way. Since transmission costs increase with throughput, it is appropriate to transfer signal without redundancy generated by the convolutional encoder. The transmission network protocols guarantee its own signal security. For these reasons, the convolutional encoder with redundant bits has been moved from the DAB server to the DAB transmitters. In each of the DAB transmitters of the SFN, the multiplex signal is encoded in accordance with the given set of parameters and then modulated, amplified, filtered, and emitted by the antenna. New data on the SFN networks with examples of implementation are included in the papers [7].

References

1. EBU, Tech 3391 "Guidelines for DAB network planning", Geneva May 2018
2. G. Faria, 'The Secret of a Successful DAB Launch? The Distributed Multiplexing', Conf. ASBU 99, Tunis, October 1999
3. "Derivation of Re-use Distance for the T-DAB allotment Plan", CEPT SE11 (94), Stockholm
4. "Criteria for the Coordination of Frequencies to be Used by Terrestrial Digital Audio Broadcasting (T-DAB) Transmitters and TV Transmitters", CEPT/ SEWG, Oporto 29 March 1993 EBU BPU 003,Technical Bases for T-DAB services network planning and compatibility with existing broadcasting Services'
5. ECC Report 49,Technical criteria of digital video broadcasting terrestrial (DVB-T) and terrestrial – Digital audio broadcasting (T-DAB) allotment planning', Copenhagen, April 2004
6. Report ITU-R BS.2384-0, 07; Annex 3: 'Social, regulatory and technical factors involved when considering a transition to DAB or DAB+ systems'
7. H. Zensen, "Real world implementations. SFN design and examples", Joint WorldDAB – ASBU DAB+ Technical Webinar Series, 2020 J. Hirigoyen, 'DAB+ network design. DAB network design', Joint WorldDAB -ASBU DAB+ Webinar Series 2020

8. T.A. Prosch, A Possible Frequency Planning Method and Related Model Calculations for the Sharing of VHF Band II between FM and DAB Systems', IEEE Trans. on Broadcasting, vol. 37, no. 2, June 1991
9. J. Doeven, 'Planning of Single Frequency Networks', ITU/EBU Workshop on Digital Broadcasting, Sofia, June 2004
10. R. Brugger, K. Mayer, RRC-06 – Technical basis and planning configurations for T-DAB and DVB-T', EBU-UER Technical Review, 2005
11. R. Rebhan, J. Zanders, 'On the Outage Probability in SFN Networks for Digital Broadcasting', IEEE Trans. On Broadcasting, vol. 39, no 4, Dec. 1993
12. R. Beuler, Optimization of digital single frequency networks', Frequenz 49, 1995
13. F. Perez-Fontan, J.M. Hernando-Rabanos, Comparison of Irregular Terrain Propagation Models for use in Digital Terrain Data Based Radiocommunication System Planning Tools', IEEE Trans. On Broadcasting, vol. 41, no. 2, June 1995
14. S. Gokhun Tanyer, T. Yucel, S. Seker, Topography Based Design of the T-DAB SFN for a Mountainous Area', IEEE Trans. On Broadcasting, vol. 43, no. 3, September 1997
15. G. Malmgren, Network Planning of Single Frequency Broadcasting Networks', Dissertation Submitted to the Royal Institute of Technology, April 1996
16. G. Malmgren, On the Performance of Single Frequency Networks in Correlated Shadow Fading', IEEE Transactions on Broadcasting, vol. 43, no. 2, June 1997
17. Ligeti, J. Zanders, Minimal Cost Coverage Planning for Single Frequency Networks' Radio Communication Systems Laboratory, Royal Institute of Technology, Stockholm, Sweden, 1997
18. G. Faria, The Secret of a Successful DAB Launch? The Distributed Multiplexing', Conf. ASBU 99, Tunis, October 1999
19. M. Ma Velez, P. Angueira, D. de la Vega, J. L. Ordiales, A. Arrinda, L-band DAB Eureka 147 field trials and coverage measurements in urban areas', IEEE Trans. On Broadcasting, vol. 48, no. 2, June 2002
20. ETSI TS 103 461 "Digital Audio Broadcasting (DAB); Domestic and in-vehicle digital radio receivers; Minimum requirements and Test specifications for technologies and products"
21. H. Zensen, Real world implementations. SFN design and examples', Joint WorldDAB – ASBU DAB+ Technical Webinar Series, 2020
22. J. Hirigoyen, DAB+ network design. DAB network design', Joint WorldDAB.- ASBU DAB+ Webinar Series, 2020

Chapter 6
Logical Channels: Layered Description of Transport in the DAB+ System

6.1 Introduction

The digital radio DAB is potentially designed to transmit potentially multimedia content. Its specification defines both organization and transmission of subchannels where programs are transmitted, as well as identifying the content of these programs.

The DAB transmitter can broadcast a number of radio programs with accompanying data and independent multimedia services. This requires a new approach to system organization. Scheduling DAB ensemble requires knowledge of:

- Information about the organization of the multiplex subchannels
- Identification of programs and data within the multiplex
- Identification of individual components of independent multimedia

DAB multiplex signal is composed of various radio programs and data. Each program can be built with different components of sound or services. The individual elements of the multiplex require unambiguous identification in receiver. It is necessary to indicate in which subchannel, in which mode, and what types of items of the program are broadcast. So, there is necessary information about the structure, divisions, and relations in subchannels of the DAB broadcast channels.

In the receiver, a variety of programs and services are obtained within a single multiplex signal. They are not distinguished by different frequencies, but by identifiers and labels that allow for easy selection of the desired content by the user. Earlier verification of types of decoders is required to check which elements of the program can be performed in an applied type of receiver.

© The Author(s), under exclusive license to Springer Nature Switzerland AG 2022
M. Oziewicz, *Digital Radio DAB+*, https://doi.org/10.1007/978-3-030-66478-7_6

6.1.1   The Control Process in the DAB+ System

The process of control in the DAB+ system is organized through the Fast Information Channel (FIC) built up of the Fast Information Blocks (FIB), where the Fast Information Groups (FIGs) are placed. Data contained in FIGs describe the frame organization and content (see Fig. 6.1).

Control over the flow of information in the DAB signal concerns – on the one hand – the organization structure of the Main Service Channel (MSC) is split into subchannels and, on the other, the organization and positions of multimedia content within these channels. Both processes are closely related. Division of channels into subchannels is adjusted to the number of currently broadcasting programs and their content, i.e. the contribution of different monomedia in multimedia programs. The control system must not only inform receiver operating system about the number and configuration of subchannels but also control the transport of various monomedia contents that make up the programs.

Depending on its functional role, there are distinguished eight types of FIG frames: three of which are intended as a reserve, while the other five types of frames are actually used: Type 0, Type 1, Type 2, Type 6, and Type 7. In each of these types detailed cathegories are defined by the code extension (Ext).

6.2   Transmission Channels in DAB+ System

The older version of the DAB system adopted four variants of the length of the OFDM symbols, and so the different intercarrier frequency spaces. Also, as a consequence, the four types of symbols formulate the four DAB frames. These four variants of the system are called *modes*. Selection of the mode for broadcasting was related to the frequency band in which the system operated.

The condition of projection 1:1 of logical onto physical DAB frames has resulted in the construction of the four modes of logical DAB frames. To distinguish the mode independent logical elements within the DAB frame, the Common Interleaved Frame (CIF) was introduced. As mentioned in Sect. 2.7.3, the CIF contains 16 frames scattered in time interleaving coder.

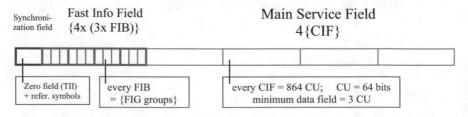

Fig. 6.1 Logical frame of the DAB+ system (mode I)

In the system DAB+, only mode 1 has been applied. It is due to the limitation of the frequency range for DAB+ in Europe to the third VHF band.

6.2.1 Organization of the DAB+ Logical Frames

Each DAB frame begins with a synchronization field consisting of null and phase reference symbols. The null symbol consists of zeros, but optionally it can include the Transmitter Identification Information (TII) codes of individual transmitters in Single Frequency Network [1]. The phase reference symbol, the same in every frame, includes the reference factors for each subcarrier. Although the phase factors are deformed during transmission, the same phase deformations relate following subcarrier factor – so the phase differences between them is equal to the modulating factor. Knowing the phase reference symbols, it is possible to start calculating the information in modulating factors on subcarriers. The next phase factors are used as reference symbols for subsequent OFDM symbol. In this way, the demodulation of frame symbols proceeds.

After synchronization field, there are located fields of the Fast Information Channel (FIC). These fields are built up from the Fast Information Blocks (FIB) – three blocks per one CIF. Each block FIB contains several groups FIG. The FIG groups contain information about the frame structure and its content in the Main Service Channel. These groups make up the Multiplex Channel Information (MCI). Further FIG groups are also possible as options, containing Service Information (SI) (see Sect. 6.6). The fields of the Fast Information Channel are processed without the time interleaving, which speeds up the transmission of FIC data in relation to other information allowing to prepare the receiver for reception. The high immunity to multipath interference for all FIG groups is secured by constant forward error correction factor equal to 1/3.

After the FIC field follow fields of the Main Service Channel (MSC) built out of Common Interleaved Frames (CIF), the basic building blocks of each logical DAB frame. In mode I, there are four CIF fields. The name CIF comes from the fact that the original frame of audio, or services, during time interleaving scattered between 16 following frames, does not go beyond the CIF field. This allows to limit the signal processing in receiver to a single CIF field to reconstruct the original audio or data frames.

The structure of the DAB+ frame in mode I is presented in Fig. 6.1.

Addressing different fields of the DAB frame and determining their capacity within every CIF frame are performed by dividing the CIF onto Capacity Units (CU). Each CU contains 64 bits. The CIF has 864 CU. Minimal field capacity within CIF contains 3 CU.

Bit capacities of various parts of the DAB+ frame are presented in Table 6.1.

Table 6.1 Bit capacity of the DAB frame

	DAB frame (bits)	Header Synchro. (bits)	Field FIC 4 x 3 x FIB (bits)	Field MSC 4 x CIF (bits)
Mode I	225,024	768	$4 \cdot 3 \cdot 256 = 3072$	$4 \cdot 55,296 = 221,184$

Table 6.2 Gross capacity of the DAB channels

Mode	Channel DAB+ Brutto (kbit/s)	Channel Synchro (kbit/s)	Channel FIC Brutto/netto (kbit/s)	Channel MSC Brutto (kbit/s)
Mode I	**2304**	**32**	**96/32**	**2304**

sub-channels are constructed of subsequent fields in following DAB frames

Fig. 6.2 The idea of subchannel organization

6.2.2   Channels and Subchannels

The DAB has the basic transmission channels with a constant structure specific to each channel:

- *Synchronization channel*
- *Fast Information Channel (FIC)*
- *Main service channel MSC*

The MSC consists of Common Interleaved Frames (CIF), each with a fixed bit rate. FIC consists of FIC frames.

Synchronization channel is divided into synchronization frames.

Frames of individual channels are connected as the DAB frame.

Throughputs of individual channels in the DAB system are presented in Table 6.2.

Individual fields of the CIF frame with appointed address are the constituent elements of the transmission subchannel of the main service channel (see Fig. 6.2).

The minimal capacity within CIF field is equal to multiples of 3CU, i.e., 192 bits. Considering the number of CIF per second (1s/24 ms), the throughput of subchannels is equal to multiples of 8 kbit/s.

6.2.2.1   Transmission of Subchannels

The main service channel is divided into transmission subchannels for programs and data transmission. In contrast to analog systems, where the various transmission

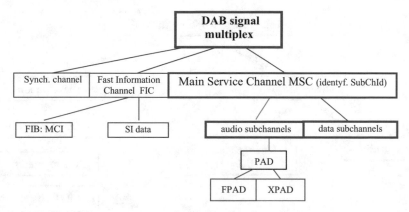

Fig. 6.3 The hierarchy of channels and subchannels in the DAB system

channels are associated with different carrier frequencies, subchannels in the DAB system are built of a subset of CU with a common address in subsequent DAB frames. This allows to turn them on or off among the other transmission subchannels by identifying address. At the same time, there may be up to 64 subchannels distinguished by their addresses.

The hierarchy of the organization of the DAB channels and subchannels is shown in Fig. 6.3.

In the Main Service Channel, in every CIF, the subchannel fields are parameterized by the starting addresses (the number of the first CU) and its capacity also in CU.

Information about the structure of multiplex configuration (MCI – Multiplex Configuration Information) is contained in the FIC channel in the FIB blocks in the FIG groups and further specified by its extensions.

6.3   Organization of Transport in the DAB+ System

Depending on the structure of data it may be carried out in one of two modes: a stream mode and a packet mode. The homogeneous data are transmitted in the stream mode in subchannels of the main service channel (MSC). Data in the packet mode can use either the Fast Information Channel (FIC) or the MSC. The capacity of packets is just 3CU, 6CU, 9CU, or 12CU, with a header containing only parameters related to the transmission (network level). So, in the transport level, demanding additionally type of content information and the address of information (e.g., in the case of conditional access), more capacious frames were introduced, so-called MSC Data Groups. For the multimedia objects, Multimedia Object Transfer (MOT) objects were introduced. To facilitate the transmission of the objects MOT, they

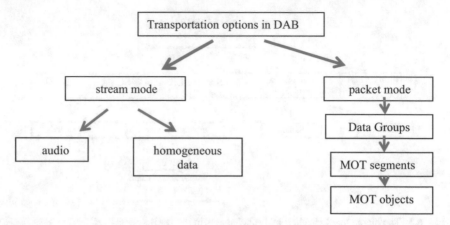

Fig. 6.4 The organization of transport in the DAB+ system

are divided into segments and, after addressing, transported as MSC Data Groups in the MSC subchannels.

Packets can be carried out in subchannels of MSC, or in Program Associate Channel PAD associated with the audio programs (Fig. 6.4).

Transport type is determined by the identifier of the transport mechanism (TMId – Trans-port Mechanism Identifier) taking the three values indicating respectively:

- The transport of streaming audio in the main channel
- The transport of streaming services in the main channel
- Packet transport in the main channel

6.3.1 Audio in the Stream Mode

Audio programs designed for transport in the streaming mode are transmitted in the subchannels of the main service channel. Since the subchannel throughputs are multiples of 8 kbit/s, which correspond to multiples of 3CU in the CIF, the application before transmission must be split into frames of such capacity. Basic capacity of the CU is 64 bits; hence, the capacity of application frames transmitted in homogeneous mode should be a multiple of 192 bits per DAB frame or adjusted to such capacity. This corresponds to a bit rate of n x 8 kbit/s.

Capacities in subchannel units CU are gross capacities including the content extended by redundant bits of code. The net throughput, corresponding to the pure content only, can be calculated considering the protection level associated with this content according to the rule

$$C(netto, kbit/s) = \{C(gross, CU)/3\} \cdot 8[kbit/s] \cdot code\ efficiency$$

In the radio DAB, the compressed signal audio is transmitted in stream mode.

Changes of subchannels throughput can therefore take place only in the increments of 8 kbit/s, what follows the sound compression system with its outer frames.

6.3.2 Data in the Stream Mode

In the stream mode, transmission in the selected subchannel is intended for one type of homogeneous service, with a fixed bit rate.

Data applications must be formatted in such a way that its gross throughput is a multiple of 8 kilobits per second (n x 8 kbit/s). This requirement is related to the minimum size of a subchannel (3 CU per CIF frame). For example, for mode I with 4 CIF fields in the DAB frame, the minimum subchannel has 4 x 3CU per DAB frame. Since there are 10.41 DAB frames per second, this means minimum subchannel throughput of 8 kbit/s ($10.41 \cdot 12 \cdot 64$ bit/s $\approx$ 8000 bit/s).

6.3.3 Packet Mode in the MSC

The packet mode in the DAB system was introduced for services:

- Of low capacity
- Requiring concurrent transmission with other services

Permissible capacities of packets in the Common Interleaved Frame (CIF) are 3CU, 6CU, 9CU, and 12 CU. This corresponds to respective subchannel bit rates equal to 8, 16, 24, and 32 kbit/s. Packets constituting one subchannel are identified by a unique address. The address comprises 10 bits, which means that the maximum number of simultaneously transported applications may not exceed 1023. An application is portioned into packets transmitted sequentially with one address. Packets of different applications (different addresses, different packet length) may be interleaved in a subchannel while maintaining the sequential order.

The general form of packet is shown in Fig. 6.5.

Packets can be transmitted in the main service channel (MSC), or in Program Associated Data (PAD) fields.

header: length, address, continuation index	service field	CRC (G_1)

Fig. 6.5 The fields of the DAB packet

6.3.4   Packet Transport in the Channel Accompanying Program PAD

6.3.4.1   Transport Layer

Value-added services in the F-PAD or its extension X-PAD (Sect. 2.4.5), packed at the transport layer in the MSC Data Groups, are divided into consecutive X-PAD frames with the type of service indicator CI (Content Indicator), if they contain one type of service. In the case of the parallel transmission of several applications, the content indicators are sent at the beginning, in isolation from the frame services [1] (Fig. 6.6).

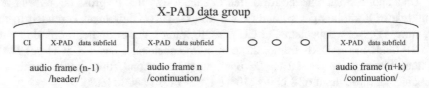

A. Groups of services of one type deployed in subsequent X-PAD /Extended PAD /frames

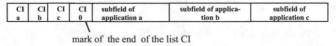

B.   Arrangement of several applications in one X-PAD field

Fig. 6.6 Methods of transmission services in the X-PAD fields [1]

6.3.4.2   The Network Layer in the PAD Channel

Division of data group on packets is natural (see Fig. 6.7).

An example of image transmission (JFIF file format) in the channel PAD is found in Fig. 6.8.

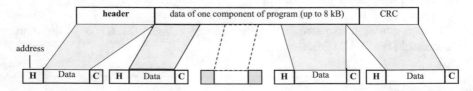

Fig. 6.7 Transfer of services in DAB packet mode. Distribution of data groups [1]

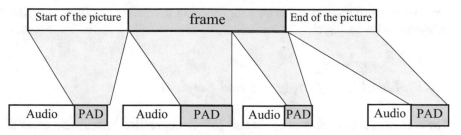

Fig. 6.8 Transport of images in the PAD channel [1]

6.3.5   Data Groups in the Main Service Channel (MSC)

MSC data groups acting as the intermediate objects for transport multimedia services are constituting the elementary addressable carrier of information in logical channels.

It is natural that the information about programs and services should appear earlier to their content and be quickly achievable when the DAB receiver is turned on. Due to the specification of broadcasting, information about the content of the programs contained in the packet headers should be transmitted independently of the data fields included in the MSC data groups. MSC data group headers contain, e.g., conditions of access to selected services.

MSC data groups carrying different data are marked with different types. So far from allocated for this purpose 16 possibilities the distinguished types of groups containing specific data items are:

- Information about the rights to use the specified services program (see Chap. 10)
- MOT object header
- Contents of the object MOT
- Conditional access parameters to the MOT object

For a closer look on the construction of applications and MSC data groups, see Sec. 7.2.2 on the segmentation of the MOT objects.

6.4   Identification of Services within the DAB Multiplex

The programs and services within the DAB multiplex signal are distinguished by addresses and numerical identifiers represented in receivers by labels. The numerical identifiers allow for quick browsing of the content on both sides: in transmitter and receiver. It is much shorter than any descriptions, so it saves the throughput of the system. However, the users require closer descriptions of the content of the available programs and services; and this is the role of labels. Labels are the short names related to the content of programs. They are used for presentation and are placed in databases in receivers outside of programs. This allows to expose them, when the receiver is turned on, without waiting for decoding the contents of all the available programs in the multiplex.

6.4.1 Identifiers of Programs and Services

The content of a single-frequency network (SFN) may originate from multiplexes of several different operators and from various countries. Therefore, the identifier of each program includes the code of country and references of multiplex. Then, each program and each of its related components within multiplex should indicate the type of the channel in which it is being broadcast: the main service channel (MSC), channel PAD affiliated with the program, and channel of services in expanded X-PAD. Indication of the transport mode, stream or packet, is also necessary.

The mentioned identifiers and parameters are contained in channel FIC in groups FIG parameterized by index and index extension: FIG(index/ext). Below briefly are described identifiers of services within multiplex.

Problems with information about other multiplexes, services of other parallel systems (DMB, FM, DRM), updating databases with their parameters, linking between different systems, and switching between systems in order to maintain the reception of selected program - are described in the specification [2].

The basic identifiers concern multiplexes, their services, and service components.

Ensemble Identifier (EId) Individual multiplexes of system DAB are identified by 16-bit code. Multiplex identifier appears at the beginning of each DAB frame in the group FIG 0/0.

Service Identifier (SId) Within a single multiplex, multiple programs are transmitted simultaneously, generally from different operators. During transmission, every program is transported in the form of fields, parts of the DAB frames. The fields of one program are distinguished by its identifier to choose, decode, and combine frames of the program at the receiver. The identifier concept has been adopted in accordance with the RDS (Radio Data System) [3]. As the programs in the DAB system can be exchanged between different radio broadcasters, or cover areas outside the framework of one country, identifier of the program includes:

- Code of the country from which program is transmitted
- Code of coverage area of broadcasting network
- Code of the program identifier

and allows to clearly distinguish each individual service. Structure of identification code of program is highlighted in Figs. 6.9 and 6.10.

b_{15} b_{12}	b_{11} b_8	b_7 b_0
country Id	service reference	service type

Fig. 6.9 The structure of the short identification code of the program [1]

b_{32} b_{25}	b_{24} b_{21}	b_{20} b_{n+1}	b_n b_0
code **ECC**	country Id	service reference	service type

Fig. 6.10 The structure of the extended SId code [1]

In the case of continuation of the known program, it is enough to limit the code to 16-bits (standard RDS code). It is a 4-bit code identifying one of up to 16 countries and 12-bit classification scheme.

New programs and new types of content are identified using the global 32-bit identifier. The country code consisting of ECC code classifying 256 groups of countries and country code in groups (as above), as well as the extended 20-bit program classification, is sketched in Fig. 6.10.

The identification code of program is included at the beginning of the header of each DAB frame in the FIG (0/2) group.

The set of service parameters (service, transport mechanism, service component description, subchannel identifier) is sketched in Fig. 6.11.

Each program may be composed of various monomedia components, up to 12. They are distinguished by identifiers of *the program component*.

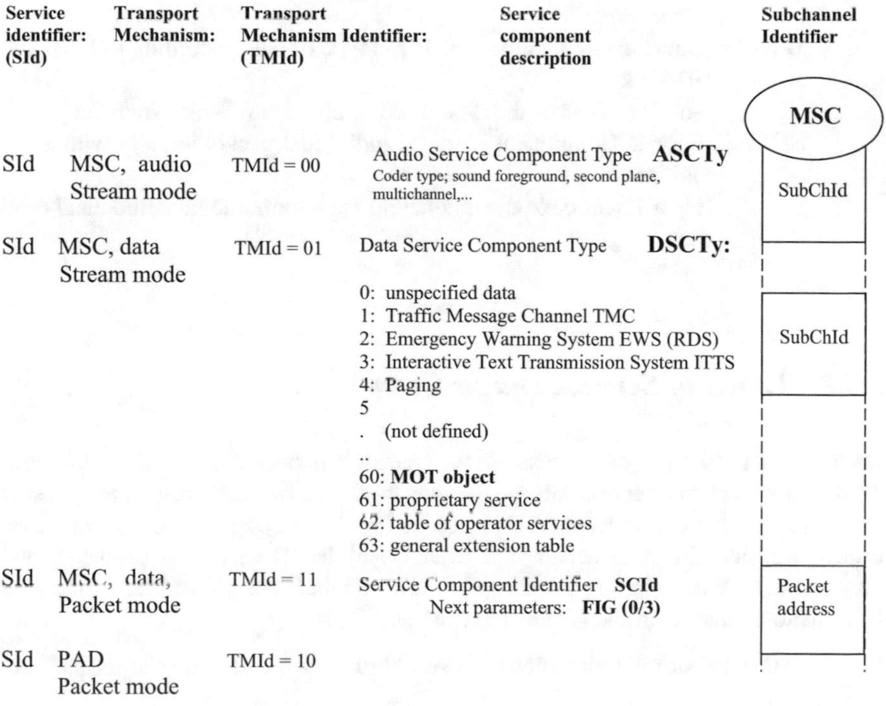

Service identifier: (SId)	Transport Mechanism:	Transport Mechanism Identifier: (TMId)	Service component description	Subchannel Identifier
				MSC
SId	MSC, audio Stream mode	TMId = 00	Audio Service Component Type **ASCTy** Coder type; sound foreground, second plane, multichannel,...	SubChId
SId	MSC, data Stream mode	TMId = 01	Data Service Component Type **DSCTy:** 0: unspecified data 1: Traffic Message Channel TMC 2: Emergency Warning System EWS (RDS) 3: Interactive Text Transmission System ITTS 4: Paging 5 . (not defined) .. 60: **MOT object** 61: proprietary service 62: table of operator services 63: general extension table	SubChId
SId	MSC, data, Packet mode	TMId = 11	Service Component Identifier **SCId** Next parameters: **FIG (0/3)**	Packet address
SId	PAD Packet mode	TMId = 10		

Fig. 6.11 Service parameters in group FIG 0

Identifiers of Services Within Multiplex

Multiplex/Ensemble Identifier	**EId** (country identifier, country reference) in FIG 0/0 at the beginning of every DAB frame
Service identification code	**SId** (Service Identifier) $\longleftrightarrow$ **PTy** (Programme Type) basic service category (FIG 0/17) for selection of content in receiver
Number of service components	**SCIdS** (Service Component Id within the Service) number of components (< 12) within the Service
For every service component	**SC** (TMId, SubChId, SCTy CA flag, A/D flag)

where:

TMId (Transport Message Identity) transport protocol

SubChId (Sub Channel Identity) sub-channel identity where service is carried

SCTy (Service Component Type) type of component: audio, homogeneous data, or packets:

> **ASCTy** (Audio Service Component Type) audio type according to ETSI TS 101 756:
> the foreground, a background; mono, stereo, sound surround, type of encoder
>
> **DSCTy** (Data Service Component Type) type of data according to ETSI TS 101 756:
> 3-bit identifier of the type of component and 3-bit extension
>
> **SCId** (Service Component Identity) individual packet identity within multiplex:
> It is a 12-bit code distinguishing each monomedia component of each program

6.4.2   Labels of Services and Its Components

Information on the programs possible for selection is presented by electronic program guide and the service labels. Service labels allow to identify the types of programs and data. The labels are placed in the databases in the DAB server on transmitter side. In the receiver, the database of labels serves to display actual program offer when starting the reception. These labels are transmitted in the Fast Information Channel in the groups of type1 and 2 [4]:

FIGs. 1,2/0: label of multiplex block (**Ensemble Id** - > literal name for the shortcut display).

FIGs. 1,2/1: programs labels (**Service Id** - > for the shortcut service type display).

The labels of service components are placed in FIGs. 1,2 /4 and data service in FIGs. 1,2 /5.

6.5   The Startup Procedures in the Receiver

The startup procedures in general situation of unknown signal are described in [2].

When the tuned multiplex is known, starting reception requires the information about the currently valid configuration of channels and programs. These features are contained in the data set MCI (Multiplex Channel Information). The MCI is a set of groups of data in the FIG type 0 with extensions from 1 to 4 and 7 in the Fast Information Channel (FIC). They describe the state of multiplex.

Similarly, in order to configure receiver to the new service, or a new channel configuration, it is necessary to transmit MCI on planned changes with adequate advance.

In the period when the multiplex configuration is unchanged, the MCI data must be also periodically transferred due to the new users turning on the signal.

Upon starting reception, the receiver input circuits carry the multiplex signal to the baseband block with the carrier frequency $f_c = B/2$ (B – the block bandwidth), where the OFDM symbols are decoded by the FFT operation; next follows differential demodulation, Viterbi convolution decoding, and the inverse frequency and time interleaving. The operating system identifies the beginning of a logical DAB frame and gradually recognizes and decodes the FIB blocks and FIG groups distinguishable by their headers. The first fully decoded group is the group FIG 0/0 containing information whether the multiplex description applies to the current, or the next configuration.

In the case of new multiplex configuration, the information on its start address is also contained in the FIG 0/0. Regardless of whether it is current or future configuration, next follows unpacking the successive FIG groups with parameters necessary to configure the receiver system.

The process of starting reception includes opening an electronic program guide (EPG; see Sect. 7.7), if such program is installed in receiver. Realization and form of the EPG system is an individual decision of each receiver manufacturers within the requirements for DAB receivers [5].

After choice of the service label, there follows selection of its subchannel address. It is contained in the group FIG 0/1.

In the group FIG 0/2 operating system finds parameters of decoders necessary for data decoding.

6.5.1   Information About the Multiplex Organization

The basic information about the multiplex organization is contained in the groups FIG type 0 with extensions 0, 1, 2, 3, 4, and 7. Relations between the FIG's of type 0 with extensions 0–4 and 7 are illustrated in Fig. 6.12. Basic data in groups FIG 0/0–7 contain:

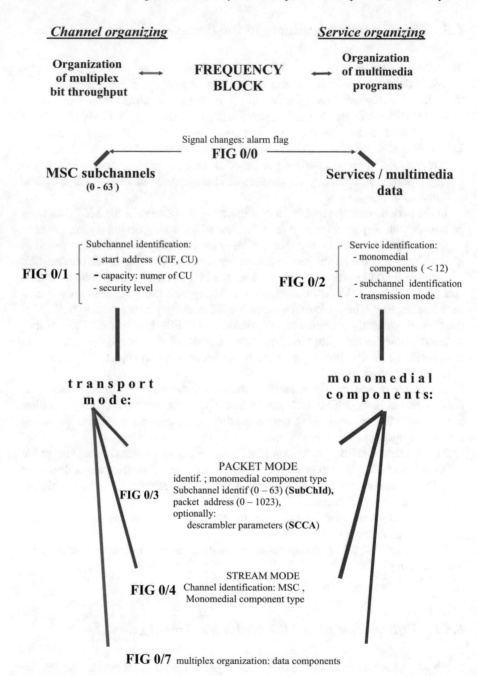

Fig. 6.12 The core organization of the multiplex: service parameters and transport modes

FIG 0/0: signaling changes at the beginning of *each* DAB frame:

- **EId** – multiplex identifier (Ensemble Identifier)
- **Change flag**:

    00 – unchanged
    01 – change of subchannel organization
    10 – change of service organization
    11 – change of subchannel and service organization .

- **Alarm flag**:

    0 – alarm messages not accessible
    1 – alarm messages accessible

- Counter of Common Interleaved Frames (CIF) for control of multiplex re-configuration
- In the case of change: number of CIF frame from which begins a new configuration

FIG 0/1: *subchannel organization* within CIF in the main service channel MSC

- Number of subchannels ($\leq$ 64).
- For each subchannel:

    - Identifier: within <0–63>
    - Start address within CIF frame (CU number: 0–863)
    - Size of subchannel fields within CIF
    - Error protection level

FIG 0/2: *organization of services* in the MSC and FIDC channels

- For each service and service component:

    - Service identifier
    - Number of components
    - For each component, the identifier of transport mechanism:

        - Audio in stream mode in MSC (type of sound, subchannel identifier)
        - Data in stream mode in MSC (data component type, subchannel identifier)
        - Packet data in MSC (service component identifier)

    - Conditional access flag

FIG 0/3: additional information on the organization of services in *the packet mode*

- For each service component in packet mode:

    - Identifier
    - Subchannel identifier

- Packet address
- Optional parameters of conditional access in SCCA field

FIG 0/4: additional information on the organization of services in *the **stream mode***

- For each component in stream mode:

  - Flag indicating service channel: MSC or FIC
  - Subchannel identifier
  - Parameters of conditional access in SCCA field

FIG (0/7): extension of the data service ***component types*** (for types outside of DSCTy)

FIG (0/8): primary service components
Parameters of the service components used by different multiplexes

6.5.2 Startup Algorithm after Changes of Service within Multiplex

The group FIG 0/0 is affixed at the beginning of each DAB frame, e.g., in mode I every 96 ms. The flag of any change in the multiplex must appear at least 6 seconds ahead of the advertised changes [1]. In this period such flag must be repeated at least 3 times. After the change, the new configuration must also stay stable for at least 6 seconds. The group FIG 0/0 includes a counter modulo 256 CIF frames. It is used for setting the starting address of new configuration, in case of changes of the multiplex organization.

As mentioned above, the CIF can contain up to 64 subchannels with a total capacity of 864 CU.

If the changes of multiplex concern the distribution of subchannels, further information about the layout of subchannels can be found in the group FIG 0/1. It includes for each subchannel:

- The identifier (6-bit number indicating one of maximum 64 subchannels)
- The start address (number of CU in the actual CIF, from which starts identified subchannel)
- For audio signal:

The index in subchannel size table (Table 8 in [1]). The net throughput of a given audio signal and the level of protection indicate in the table the gross size of frames in CU of adequate subchannel in CIF, marked by the index

- For other signals:
- The protection level
- The subchannel frame size (number of consecutive CU's assigned to a given subchannel in CIF)

If the changes affect the organization of multimedia in the MSC channel, further details are provided in the group FIG 0/2. They include for each service:

- The service identifier
- The identifier of Access Control System
- The number and parameters of each multimedia component together with its type and identification of subchannel in which this component is transmitted

The number of multimedia components of one service may not exceed 12. The number of different types of monomedia, from which components of the service can be chosen, is up to 1023.

Depending on whether the component is transmitted in stream mode or packet mode, further information is contained in group FIG 0/3 or FIG 0/4.

Additional information about the organization of services in the stream mode in group FIG 0/4 indicate – for each monomedia constituent from FIG 0/2 – whether the transmission subchannel is in the channel MSC, as well as the parameters of the conditional access (descrambler) for this subchannel.

Organization of multimedia in packet mode with additional information is included in the group FIG 0/3 and comprises an identifier of each multimedia component from FIG 0/2, the index of the conditional access, subchannel identifier, and the address of the packet with the parameters of the descrambler.

Frames of FIG type 0 with extensions 0–4 are transmitted in the first block FIB in each DAB frame.

6.5.3   Reconfiguration of Subchannels and Services

Due to the correlation of the throughput of source encoder, the efficiency of the convolutional encoder, and capacity of subchannels, the change in subchannel throughput requires an advance of at least 15 frames, i.e., 384 ms, for the steering information. It is associated with the reorganization of the time interleaver at the transmitter (Sect. 2.7.3).

Any change in the structure of the broadcast service or new division of subchannels on the transmitter side must be communicated in advance to the buffer of the receiver to adjust its parameters and proper decoders and to act in accordance with the new division of subchannels.

Information about the organization and structure of multiplexer and services is transmitted in the Fast Information Channel (FIC). At a transmitter, the FIC frames do not undergo time interleaving and alike in the receiver do not pass the inverse operation. Faster information allows to reconfigure the parameters of the buffer in terminal. Similarly, faster information on reorganization of services allows for updating the database of service labels for presentation of the new changed programs.

6.6   Optional Parameters of the Programs and Services

The data in the Multiplex Configuration Information (MCI) group, necessary to activate the receiver, must necessarily be transmitted by each DAB transmitter, and each DAB receiver must be able to decode the MCI. Digital radio system limited to such basic functions – from the point of view of the user – does not differ much in its program offer from an analog radio system. However, the possibilities of digital transmission and the digital signal processing allow for a significant expansion of this DAB service offerings. Developing such additional offer requires the use of additional data and parameters described in the standard system [1] as the Service Information (SI) and discussed in detail in the Specification [6]. Using this supplementary information requires expansion of the operation system on both the transmitting and receiving sides and is left to the decision of the administration and manufacturers as an optional offer.

Service Information (SI) is transmitted in the Fast Information Channel (FIC) in the Fast Information Groups (FIG). These data concern the additional parameters and mechanisms of the current multiplex, or other multiplexes in other blocks of spectrum, as well as parallel programs broadcast in systems AM and FM.

Detailed information about the types of monomedia components, applied languages, the access conditions, relationships between programs, etc. is placed in groups FIG of type 0 with extensions 5–31.

FIG group of types 1 and 2 contain service labels mentioned in Sect. 6.4.2.
FIG groups of type 3 and 4 and 5 are reserved for future use.
FIGs of type 6 contain information on conditional access; see Appendix D.
FIG group 7 serves as the end marker in FIBs not completely filled with FIGs.

6.6.1   Multiplex Parameters Described in the FIG Groups of Type 0

Databases including optional parameters are contained in the FIG groups of type 0 with extensions 5, 6, 9, 10, 13, 14, 18–19, and 24–26. The length of each of these groups is a free parameter. Flag C/N (current/ next), in the case of transmission of databases specified by the signaling SIV (Service Information Version), indicates that the group of the given extension includes the beginning of the base (0) or another portion of the data (1). In case of any changes to the database, the key CEI (Change Event Indication) is used individually for each database.

Groups of type 0 with extensions 0–4, 7, and 8 include discussed earlier multiplex channel information MCI and supporting data.

The content of individual groups is discussed in detail in the specification of the system [1], and the introduction to its implementation [5–7], where information on the repetition frequency of these groups and their links with other services is placed also.

In the following are given some basic information about the content of optional parameters of FIGs of type 0:

FIG 0/5:   Information on the language of each audio component of the service (Service Component Language); language codes from ETS 300250, annex 1 of part 5

FIG 0/6:   Information on the relationship of the services (service linking information)

FIG 0/9:   Local time offset (LTO Country and International Table)
Local time offset and Extended Country Code (ECC)

FIG 0/10: Date and Time
Timestamps synchronized with the received signal, optionally in the zero symbol of the DAB frame, in the synchronization channel. Date by modified Julian calendar, universal time to the nearest hour and minute, or in an extended version additionally seconds and milliseconds; the country code; local time

FIG 0/13: Information about the application type. Link between application and the appropriate decoder

FIG 0/14: Parameters of Reed-Solomon code applied in subchannels

FIG 0/17: Program Type (Pty)

FIG 0/18: The base of verbal messages (Announcement support)
The mechanism for the introduction of verbal messages through the interruption of the current program. The messages are parameterized by:
- Identifier of the source (SId)
- Classification of the message types (selection flags of type Asu)
- Identifiers of clusters performing the role of the mailboxes. In the multiplex can be located up to 255 clusters
Clusters determine the levels of interrupt of messages. Established international categories of message types are defined in the table

Bit in flag field	Type of message
b_0	alarm
b_1	organization of traffic
b_2	organization of transport
b_3	warnings
b_4	message digest
b_5	local weather
b_6	announcement about the event
b_7	special event
b_8–b_{15}	reserved to define

Cluster "1111 1111" is reserved for alarm, authorized for the interruption of any program.

FIG 0/19: Switching the verbal messages (Announcement switching)
Parameters of messages allowed to interrupt the service:
- Identifier of the cluster (of which mailbox)
- An indication of the types of messages allowed to interrupt service (flag Asw)
- Identifier of the subchannel, where message is placed
- Optionally: definition of the region to which is limited the message reception

FIG 0/20: Service Component Information
Determination of the start time or the end time of the specified services for a specified group of receivers

FIG 0/21: info. on frequencies of services within the block (DAB multiplexes + FM)

FIG 0/24: offer of programs in the parallel DAB multiplexes

FIG 0/25: information about messages in the parallel multiplexes

FIG 0/26: Information about switching messages in other multiplexes
Features of blocks of other programs (OE = 1)

6.6.2   Selected Software Services

A. Announcements

Announcements are verbal short messages placed in the subchannels of the main service channel (MSC). It could be weather reports, stack market data, information on sport, road, transport and alarm communications. Programs authorized to receive the indicated type of announcements form the so-called announcement cluster. The set of authorizations of various announcements is transmitted in groups FIG 0/18, where in the field of announcement support (ASu) particular flags indicate the types of announcements authorized to interrupt the program. The FIG 0/18 frame also contains the identifiers of cluster that allow the announcement.

When the announcement is in another subchannel of the tuned multiplex, the programs authorized to receive the indicated message are informed about the address of the announcement subchannel in the group FIG 0/19. In the field of announcement switching (ASw), the announcement type is marked with an appropriate flag.

If the transmission is not switched off by the user, the reception is redirected to the subchannel with the announcement and – when it is finished – returns to the original subchannel. Return to the original channel is determined by resetting of the Asw field.

Schematically, the handling of announcement algorithm is depicted below (Fig. 6.13).

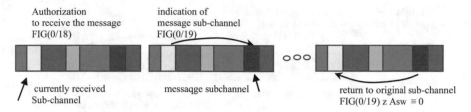

Fig 6.13 Reception of the announcement message

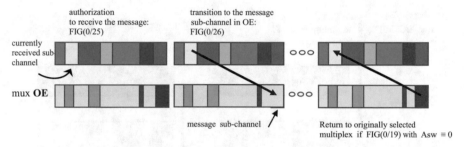

Fig. 6.14 A passage to announcement in Other Ensemble

When the content of the announcement is in another multiplex (OE – Other Ensemble), the authorizations are determined in the groups FIG 0/25, and the index of switching on the announcement multiplex is in the group FIG 0/26. In this case, it is necessary to tune the receiver to the frequency of the multiplex accommodating the message. After reception of the message appears the FIG 0/26 with field Asw $\equiv$ 0. It indicates that the receiver must retune to the original multiplex, as indicated below (Fig. 6.14).

A0. ALARM Signals – Warning Messages

A special case of announcements are warning messages informing about the threats to the population within the range of a given transmitter – these are alarm announcements.

To reach a multiplex listener independently of the actually receiving service, an alarm message is announced in the first Group of Fast Information FIG 0/0 in the Alarm flag field. For such a signal, the receiver reads the address of the subchannel of the alarm from the FIG 0/19 and switches the receiver to this subchannel when it is in the tuned multiplex (Fig. 6.15).

When the subchannel of the alarm is in another multiplex, its multiplex identifier is read in the group FIG 0/26, and the receiver is tuned on this multiplex. The address of warning data subchannel is received as above in FIG 0/19. When warning message is received, reception is retuned to the original multiplex – and the starting algorithm applied (Fig. 6.16).

These operations are repeated over the entire duration of the warning time.

ALARM flag = 1   address of sub-channel
in group FIG(0/0)   with warning message
                           in FIG(0/19)

FIG(0/0)

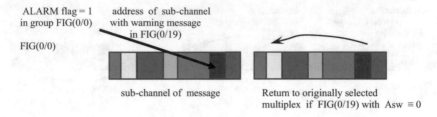

sub-channel of message   Return to originally selected
                                multiplex if FIG(0/19) with Asw ≡ 0

Fig. 6.15 The alarm message in the tuned multiplex

ALARM flag =1   FIG(0/26): re-tuning
in group FIG(0/0)   to multiplex EId
                    with alarm message

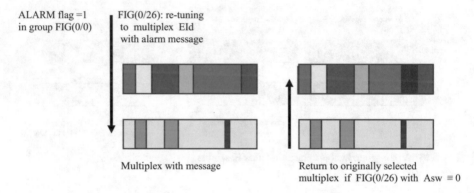

Multiplex with message   Return to originally selected
                                multiplex if FIG(0/26) with Asw ≡ 0

Fig. 6.16 Alarm message in another multiplex (Ensemble)

B. **Linking Information**

The linking information allows to continue the selected program also out of the range of transmitters of the tuned multiplex. This is the so-called hard linking. When this option is not present, because no other radio station is suitable, the system allows you to choose a program with a similar category in other multiplexes, and when such options do not appear, the next step is searching alike programs in FM or AM signals. It is so-called soft linking.

Identification of an identical (hard) or similar (soft) program is achieved by comparing the identifiers of the current and other programs (the linkage set) in the current or neighboring coverage areas with the signal. The receiver system scans the program identifiers placed by the operator in the bases FIG 0/6 for a given multiplex and FIG 0/24 for parallel multiplexes and selects the optimum according to the criteria: availability and signal strength.

The programs that make up the related cluster are identified by the Linking Set Number (LSN), the international coverage index ILS (International Linkage Set): domestic - outside, and the identifier list Identifier List Quantifier (IdLQ). Type of the program identifier "Id list" depends on the type of system (OFDM, FM, DRM, AM) according to the IdLQ flag.

Choosing an adequate database enables a set of parameters (flags) constituting the key to the database you are looking for. The database whose data does not fit into one

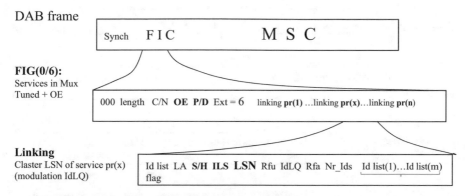

Fig. 6.17 Information on linking information within DAB frame

FIG 0/6 group is transmitted in successive groups of this category, which is indicated in the group header.

The transition algorithm for the associated program consists the following steps:

The current program → cluster number of associated LSN → associated program identifiers → selection → selection of frequency from FI base in group FIG 0/21

The structure of the group FIG 0/6 is shown below. The data indicator (database key) is created by collection of flags marked in bold (Fig. 6.17).

Acronyms

relations pr (1) – links of the current program
Id list flag – list presence flag (0, missing; 1, present)
LA (Linkage Actuator) – current or future list
S/H (Soft/Hard) – similar or identical program
ILS (International Linkage Set indicator) – the area of domestic coverage or abroad
LSN (Linking Set Number) – the cluster number of linked programs
IdLQ (Identifier List Quantifier) – modulation system identifier → type of program
    identifier

C. Frequency Information

Information on carrier frequencies of radio signals of systems DAB, DAB +, DRM, FM with RDS, and AMSS:

– In the coverage area of the currently received signal
– In adjacent areas

are included in the database in the fast information groups FIG 0/21, one or the followings, as indicated in the C/N flag (0, start; 1, continuation), in the group header.

Carrier frequencies are combined into lists for technically different systems defined by the *R&M* indicator (range and modulation) and distinguished by an id

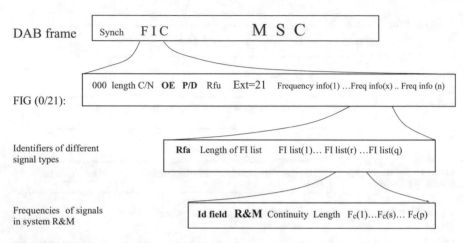

DAB frame

FIG (0/21):

Identifiers of different
signal types

Frequencies of signals
in system R&M

Fig. 6.18 Frequency information within the DAB frame

field identifier indicating, e.g., a specific multiplex, DRM program, RDS code, or AMSS program identifier.

The general structure of the FIG 0/21 frame is sketched in Fig. 6.18.

The fat font indicates the frequency base key.

The ID field depends on the system (OFDM, FM, DRM, or AM) according to the R&M flag, which defines the system. The identifier field indicates a specific signal. The continuity field precedes the possible interruption during the switching of signals from current to linked.

Carrier frequencies (Fc) are included in the ranges depending on the system.

On the transmitting side, the frequencies of multiplexes and other radio systems in their coverage area are placed into bases in FIG 0/21.

On the receiving side, the system scans the program identifiers and respective carrier frequencies from this database.

D. **Identification of Current Services in Tuned Multiplex**

Information about current services in received or other multiplexes or systems (FM, DRM, AM) is transmitted in the databases of the Fast Information Group FIG 0/24. For data related to the tuned multiplex, the OE indicator equals 0 and 1 for the remaining signals.

This data allows to choose appropriate program on the receiving side.

The FIG 0/24 frames contain lists of multiplex identifiers including the indicated program with the SId identifier (from the received multiplex).

The database key is marked by flags given in bold. The program identifier SId contains a 32-bit description for the beginning or continuation of the database and 16-bit for the data change.

Information about changes in the database is transmitted via the Change Event Indication (CEI) flag. The change of the entire version is announced via the SIV (Service Information Version) field (Fig. 6.19).

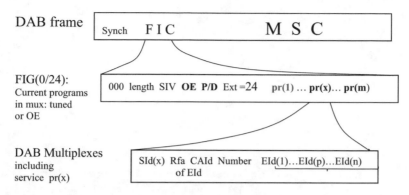

Fig. 6.19 Information on actual programs within the DAB frame

Acronyms

SIV – Service Information Version
SId – Service Identifier
CAId – Condition Access Identifier
EId – Ensemble Identifier

References

1. ETSI EN 300 401 "Radio broadcast systems: Digital Audio Broadcasting (DAB) to mobile, portable and fixed receivers"
2. ETSI TS 103 176 "Digital Audio Broadcasting (DAB); Rules of implementation; Service information features"
3. CENELEC EN 62106, "Specification of the radio data system (RDS) for VHF/FM sound broadcasting in the frequency range from 87,5 to 108,0 MHz"
4. ETSI TS 102 980 "Digital Audio Broadcasting (DAB); Dynamic Label Plus (DL Plus); Application specification"
5. 'Digital Radio Receiver Profiles', World DMB Forum, December 2008
6. ETSI TR 101 496–1 "Digital Audio Broadcasting (DAB); Guides and rules for implementation and operation; Part 1: System outline"
7. ETSI TR 101 496–2 "Digital Audio Broadcasting (DAB); Guides and rules for implementation and operation; Part 2: System features"

Chapter 7
Multimedia Applications: Protocol MOT

7.1 The Development of the Concept of "Value-Added Services" in the DAB System

The DAB system is known as radio because of its primary purpose. However, the capabilities of the system are much wider. The DAB system is adapted to transmit multimedia objects.

Multimedia objects are objects consisting of many (multi) media such as text, graphics, images, audio, and video. The various media related in the logical unit, synchronized in time and spatially, create multimedia objects.

A special class of media is known as hypermedia. Hypermedia form a network of information elements called nodes, which are linked relationally in a way that allows non-sequential dialing them.

DAB terminal adapted to receive the media must have software that allows decoding and presentation of selected media. In practice this means that the DAB receiver is approaching in their construction the specialized computer. In the case of notebooks or laptops – with the appropriate PCMCIA card. In case of mobile phones – radio system coupled with the screen with exposure required for video resolution.

7.2 Multimedia in Broadcasting Systems

To transport media in digital broadcasting DAB channels independent of their individual characteristics (synchronization system, frame organization, selection and distribution, etc.), a uniform system of transportation of multimedia objects was introduced. It is the Multimedia Object Transfer (MOT) protocol [1]. Objects MOT play the role of universal capsules allowing to transmit any media via DAB channel and – after being transported to the receiver – to use their own parameters to unpack, merge in the radio buffer, decode, and display.

Each object MOT consists of a header containing information about the object and the field locating transmitted object. Due to the nature of a broadcast system, the header and object field are transmitted independently. In order to quickly organize the object reception, the user should have the shortest possible time to access offer of broadcasted programs and services after turning on the receiver. So, the header is transmitted more frequently than the media objects themselves. Considering the possibility of errors during transmission, which requires repetition of damaged portions, both header and object field are divided into segments, thereby reducing repetition only to the damaged segments, not the whole object.

The organization and rules of operation of protocol MOT are described in the Specifications [1, 2].

7.2.1   Protocol of the Multimedia Objects Transfer (MOT)

To expose the received multimedia object in the DAB terminal, an operating system of the receiver must guide it to the right executive program(s). Information about the type of media object is contained in the frame header of the MOT object. In addition to information about the type of object, the MOT header contains the conditions and parameters of its presentation at the receiver.

MOT object consists of the following main parts (Fig. 7.1):

– Header core
– Header extension
– Object body (up to 256 MB)

The *core header* specifies:

• The length of the header and object fields
• The type of object indicating the category (e.g., general data, text, image, audio, video, MOT object, the reservation of space for further use)
• The specific subtype of the object (type of encoder or file format)

The *header extension* contains the characteristics and parameters determining exposure conditions of the application of object MOT. It includes:

(a) The time parameters defining the conditions for the presentation of the object:

  – The date of creation of applications (CreationTime).
  – The start of the exposure period (StartValidity).
  – Expire of exposure time (ExpireTime).
  – Start time of exposition (TriggerTime).
  – The period of repetition (RetransmissionDistance). This is the maximum time gap between successive repetitions of transmission of MOT object reported with the resolution up to the 1/10 second.

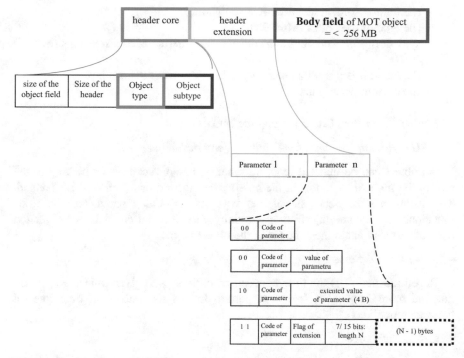

Fig. 7.1 Organization of the MOT frame

The meaning of the five of time parameters of the object MOT exposure, specified in the extension header, is illustrated in Fig. 7.2. All timing parameters are given in UTC format (Universal Time Coordinated). It determines the time with an accuracy of milliseconds (longer form) or minutes (shorter form). The time frame also includes a date administered according to the Julian calendar MJD (Modified Julian Datc).

(b) The priority and the version of the object
(c) The parameters identifying the body object;

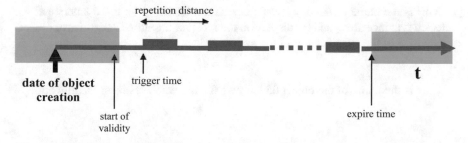

Fig. 7.2 Illustration of time parameters of MOT object in extension of the frame header

- The unique content identifier (ContentName).
- The type of the object (MimeType)
  (type of object according to Multi-purpose Internet Mail Extensions (MIME) [3–7]).
- The label. It is a short alphanumeric description for the display.
- The content description.

Further Parameters Characterizing the MOT Object

– *The Version Number of the Object (VersionNumber)*

An object can be updated (e.g., information about vacancies of parking in the city or the number of rooms in the hotel) when the exposure remains unchanged. To distinguish between successive versions of data, parameter "version" (VersionNumber) varying from 1 modulo 256 was introduced. New version demands to read again the contents of the object field.

– *Repetition Distance*

Repetition period is the time between repetitions of a given object. Repetition may be more frequent – repetition period defines the maximum time interval between repetitions.

– *Group Reference*

Each multimedia object consists of homogeneous objects (monomedia) logically related to each other. The individual components constitute individual objects MOT identified by common parameter "group identifier" (GroupID) forming the first part of the "group reference." The second part of the "group" indicates the numbers of items. Specification allows up to 65,535 of the object components.

– *Priority*

Parameter taking the value from 255 (lowest priority) to 0 (highest priority). In the case of overloading the memory of the terminal, it is priority which determines the order of keeping individual objects MOT. This also applies to components of a multimedia object.

– *Label*

This is the name of the object category with a length of up to 16 letters and the selected typeface, designed for display on the receiver monitor. Categories of objects are defined in the specification of the system [8].

– *ContentName*

This is the name of the object that allows for the detailed object identification

– *Content Description*

Text describing the contents of the object and intended for display on the receiver monitor. Text length is determined by the indicator of non-standard length of the header extension.

– *Application Specific*

This parameter is used to pass data between the server and the DAB transmitter and is determined by the system operator as appropriate.

The algorithm of the basic structure of the MOT objects is presented in Fig. 7.3.

7.2.2   Segmentation of MOT Objects in the DAB System: Data Groups

The body field of the MOT object has a capacity of up to 256 MBytes. The transmission of such file could block the DAB channel for other objects for at least 2 minutes, even for the lowest level of protection.

On the other hand, even a single error of MOT transmission would force repetition of the entire object.

Due to the indicated reasons, the MOT objects are transmitted when divided into smaller pieces. In the transport layer, the MOT objects are divided into at least two segments with a capacity of not more than 8191 B (2^{13} bytes) each. Segmented header (core and extension) and the body field are divided separately. The size of segments of the header may be different from the size of the object body segments. The process of division and distribution of MOT objects is shown in Fig. 7.4.

Each segmentation header is composed of two parts:

– *Repeating counter*, which includes repetitions left to perform. It is assumed that the repetition should last up to object expire time.
– *Segment length*, this field describes the segment length in bytes.

The segment header does not contain information about the transport mode of segments. Therefore, each segment, together with the header, is wrapped up in an extra double header and control error creating a data group.

Data Group is specified by:

– Data Group header
– Session header

Data Group header ensures continuity of all data segments and specifies the type of segments, the presence of conditional access, and the presence of the session header.

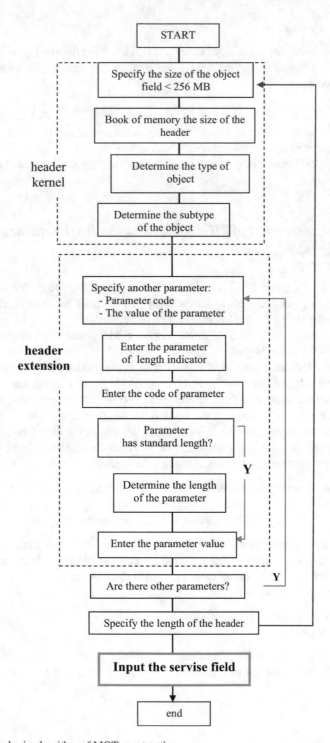

Fig. 7.3 The basic algorithm of MOT construction

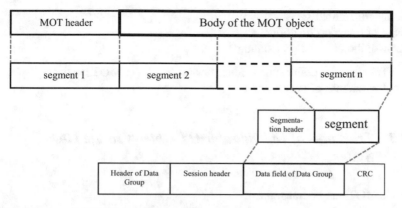

Fig. 7.4 Division of MOT object into segments and projection onto Data Groups (after [2])

Data Group header fields include:

- Flags:

  - Extension header field
  - CRC code
  - Field of the session
  - Fields of the user access
    (The last two fields determine the occurrence of the session header.)

- The type of object in the Data Group packets:

  - MOT header segment
  - MOT body segment
  - The parameters of the conditional access CA [9]
  - Other data

- *Continuation index* (modulo 27). Continuation index is incremented only if the queue of data groups is interrupted by a group of different content.
- *Repetition index*. It is a number of repetitions of the group, a number from 0 to 26. Number 27 (1111) represents the continuity of repetition.
- *Field extension*. This field contains the frame of conditional access (DGCA) with the field IM (Initialization Modifier) and additional parameters CA (see [9] on conditional access)

Session header includes:

- An indicator of the final segment
- The segment number from 0 up to a $32,767 = 2^{15}-1$, which corresponds to the body of maximum capacity 256 MB, because each segment contains up to 8 kB
- The field of user access

and optionally in the field of user access:

– 3 bits for future use
– Flag of transport identifier
– Length indicator field including:

  – Transport identifier (this is identifier of the object MOT header)
  – The address of the end user

7.2.3 Transport of Individual MOT Objects in the DAB System

Objects MOT can be transmitted:

• In the main service channel (MSC) in packet mode
• In the channel PAD of services associated with the program

Transportation of header segments is subject to different rules than the transport of body segments:

(a) Header must be transmitted at least once before the broadcast of the service.
(b) Header is transmitted frequently enough for the terminal user to get quick access to information about the range of services offered by the operator.

Segments and Data Groups of one MOT object must be transmitted sequentially, along with the increasing number of segments.

7.3 Transmission of Slides in the DAB Channel

To enable presentation of slides or a sequence of images in the receivers equipped with monitors, specification of the slideshow that uses protocol MOT (MOT SlideShow) [10] was developed in the DAB system. Applications of this specification include a photo illustration of appointed messages, photo stills of song performers, or program independent presentation of the advertised services. Broadcasting the MOT SlideShow is announced in the group FIG 0/13.

In the DAB system, two profiles of slideshow have been proposed:

• *Simple profile* (limited range of colors; images smaller than the screen are placed in the center with the black border; higher ones are cut at bottom)
• *Enhanced profile* (greater range of colors, on the receiving side the possibility of scaling dimensions in the range of 0.5–1.5, control of the rate of the slideshow)

Slideshow requires a receiver equipped with a screen display of recommended dimensions of 320×240 pixels; other dimensions may cause deformation of the slide (for the enhanced profile, minimum size is 160×120 pixels). Required decoder of images in JPEG or PNG system requires a resolution of:

- Up to 8 bits for each of the four colors for a simple profile
- 15 bits per pixel for enhanced profile

Animations of PNG presentations (APNG specification) with a duration of one frame not less than 100 ms are presented at a rate no greater than 10 frames per second. If the processing speed of the receiver will not be sufficient, in place of the animation, there will be presentation of selected stationary frames.

The order and time of exposure are controlled by the operator of the slideshow on the transmitter side using the parameter of time viewing (TriggerTime) transported along with the slides.

7.3.1   Slide Transportation

Slides are transmitted to the receiver successively, each slide with its own parameters as the individual media object MOT. For the purpose of transmission, the object files are divided into segments and packed in a Data Groups transported alternatively:

- In the PAD channel as a service accompanying audio program
- In the MSC channel, in the packet mode as a component of service
- In the MSC channel, in the packet mode as a basic service

If slides are intended as a presentation illustrating the program, the slides can be sent either in the channel accompanying program (PAD) or as a component of the program in the Main Service Channel. As a service independent of the program with its own identifier slideshow will be transported in the Main Service Channel (MSC). This function can be served by slides for the purpose of advertising services, promotion of goods, or imaging events regardless of the current program.

To enhance the credibility of the correct slide reception, the repetitions of segments can be used according to the protocol of broadcasting objects MOT, but without interleaving segments of different slides. Individual slides are unpacked, merged, and decoded in time depending on parameters of the receiver processor. To start their exposure consistently with the illustrated program, the exposition moments are determined by the commands with the parameter TriggerTime, processed independently. Such procedure requires a memory management with buffers serving the individual processing steps, in accordance with Fig. 7.5a. In the case of the enhanced profile, the buffer of integrated MOT object is multiplied, as shown in Fig. 7.5b, in order to be able to collect a larger number of slides (in simple profile new slide removes the previous one from the monitor). Thanks to this process, the exposure of slides in the receiver can be respectively adjusted.

In the simple profile, the firing of an image follows its configuration (rendering finally ending as the bitmap). In the enhanced profile, it is assumed that the image formation takes place in real time, and thus the firing of the slide can be actuated in the MOT object buffer, from where, after a short delay on the formation, the slide is

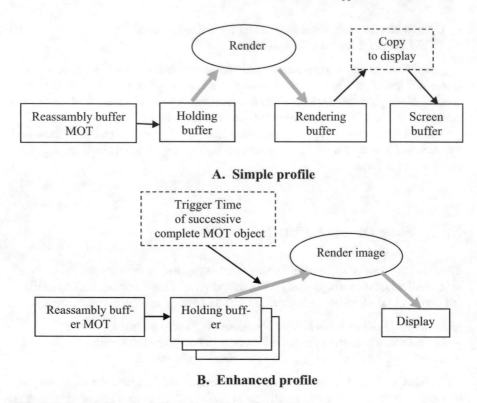

Fig. 7.5 Stages of decoding slides in the DAB receiver [10]

displayed on the screen. If the buffer is full, previous MOT objects are gradually eliminated, until there appears the space for the current slides.

Capacity of the remaining buffers is determined by the capacity of the largest MOT object and the size of an image.

7.3.2   *Parameters for the MOT SlideShow*

Exposition time of each slide ("Trigger Time") is typically presented in the MOT header extension of object that contains the slide. Exposition time is synchronized with the reference time included in the group FIG 0/10. However, the slide can be transported to the buffer of MOT objects at the receiver without this parameter and the adequate time of exposition passed later in the header ("Header update") with indication of the corresponding object by the content parameter ("ContentName").

In the case of the enhanced profile, there may be a series of updates for MOT slide objects waiting in the buffer.

7.4   Transmission of Audio-Visual Objects: Video

Each audio-visual message consists of succeeding scenes, composed of various interactive, audio-visual objects:

- Video
- Audio
- Text
- Graphics two- and three-dimensional
- Synthetic music
- Synthetic sound objects

Transmission of such scenes requires not only the coding of individual objects but also their mutual spatial and temporal relations. To describe these relations in the DAB system [11], a subset of relations described in the specification of multimedia objects [12] was selected and adopted.

In order to reconstruct complex multimedia objects of one scene, there is required the decoded description of the individual monomedia and a set of additional parameters allowing to compose these components back into one scene. Such parameters are called scene descriptors. The descriptors are indicating the spatial and temporal coordinates of the individual components of the encoded scene. Relations between the components determine the scene composition. The transition from the representation specified by the descriptors of the particular composition is the interpretation.

The specification [12] allowed to choose for the DAB/DAB+ system the information and transmission procedures of multimedia objects:

- Model of management of the terminal buffers and its timing relationships
- Encoded representations of the time-space audio description of visual scenes (Binary Format for Scenes – BIFS)
- Encoded representations of the identification and description of audiovisual streams with logical relationships between the various sources of information (Object Descriptors – OD)
- Encoded representations about synchronization (Synchronization Layer – SL)
- Representation of the common multiplex of individual streams

The relationships between the basic concepts of the audio-visual objects are shown in Fig. 7.6. It shows schematically the creation of elementary streams: the description of the scene separately and apart the coded descriptions of individual objects, video, and audio. These streams are combined in a video multiplexer, whence – through the Reed-Solomon encoder and the selected transmission system – are transmitted to the receiver.

The scene description does not relate directly to the elementary streams but uses the descriptors of individual objects. Development of procedures in each of these independent areas can therefore be considered without the need to reconstruct the whole method.

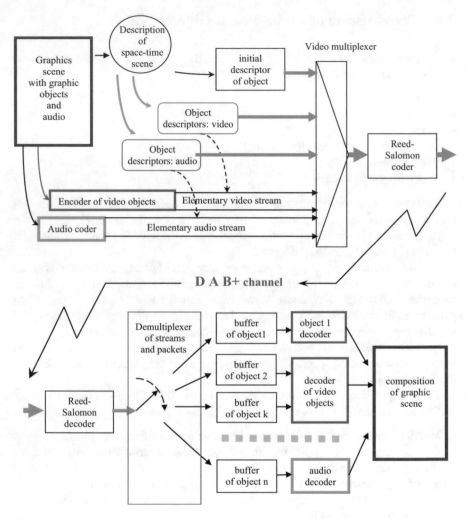

Fig. 7.6 The structure and transmission of typical example of video (According to [10])

In the process of transmission, the parameters synchronizing and locating the objects of the scene should precede the codes of the objects themselves

The objects of the scene, after passing relevant encoders, outcome in the form of packages supplemented by various synchronization parameters, such as time references for the time hierarchy of objects and time stamps specifying the recommended moments of decoding and composing the scene in the receiver.

The output frames of the encoder of individual objects are combined into elementary streams divided into packets on the synchronization layer after turning on information on synchronization and temporal relations and thus creating packetized elementary streams. Similarly, object descriptors that describe the scene after encoding are placed in packages and multiplexed into a stream containing parallel parameters linking the objects in the scene with its synchronization (OD/BIFS – Objects Descriptor/Binary Format for Scene).

Flags and indicators in headers of packet of synchronization layer refer to:

- The beginning of the unit
- The end of the unit
- Conditional access (yes/no)
- Timing parameters
- A common reference time for synchronization of objects forming the stage

The multiplexed data stream is divided into access units (AU) with time stamps attached for the time synchronization of scenes of video at receiver.

Depending on the specifics of the transmission system, the access units (AU) during transportation can be further subdivided and rebuilt after crossing the channel in the buffers of the individual streams at the output of video demultiplexer.

Signal processing in the receiver puts demands on the size of the buffer and a process for its management, so as not to overload it. These requirements are announced at the beginning of the session by the elementary stream descriptors to decide whether it will be possible to decode a particular session in a particular type of receiver.

7.4.1 Video Transmission in the DAB System

Packages of individual streams (PES – Packetized Elementary Streams) identified by indicators PID (Packet IDentifier) and time synchronized are combined into complex frames called containers in accordance with the transport standard MPEG-2 TS ([13], part 1, Systems).

A stream of containers from video multiplexer after passing the RS encoder is then projected onto symbols of a stream mode of DAB system in the main transmission channel MSC according to Fig. 7.7.

After transmission in the DAB receiver, video signal is processed in reverse order. The recreated containers after passing DAB channel are then divided into individual synchronized streams by selecting and combining packets with PID index for each stream. The individual streams are loaded into the relevant decoders and – when synchronized and composed according to the parameterization of descriptors in monitor buffer – displayed on the screen.

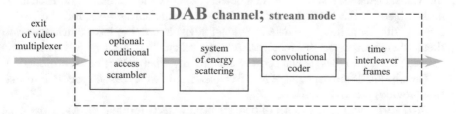

Fig. 7.7 Shaping the video stream at the DAB transmitter

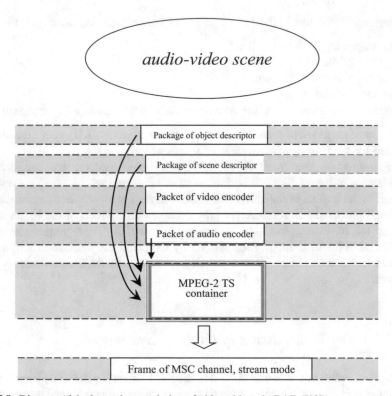

Fig. 7.8 Diagram of the layered transmission of video objects in DAB ([13])

Layered description of transmission of media objects can be represented as the graph in Fig. 7.8 [11, 13].

7.5   Carousel of MOT Objects: MOT Directory

The MOT carousel [2] is created by a set of files formatted as the MOT objects, on the transmitter side in the server of MOT applications, and periodically retransmitted to expose in the DAB terminals with the possibility of repetition and updating the individual elements. Example of such objects is a collection of messages, each as an audio object in the MOT format, regularly retransmitted as a sequence of MOT objects with the ability to update individual news, in order to have quick access to them when turning on the receiver. Another example is a set of advertisements formatted as audio files, actual sport results, or any other collection of information.

Description of the contents of MOT carousel and fast access to specific information provides the MOT directory.

MOT directory is a file of set of headers from objects forming the MOT carousel along with the parameters defining the terms of their presentation. Reception of file containing MOT directory allows for selection of any object from carousel and starting procedures for this file presentation, even if the transmission of the MOT carousel has not yet occurred. In the receiver, there is always only one MOT directory: a possible new directory replaces the previous one.

The length of the directory depends on the length of the list of parameters and the number of advertised objects in the MOT carousel. The directory header contains the following basic data:

– The length of the directory.
– The number of reported objects.
– The period of repetition of carousel objects.
– The length of the segments after splitting directory during transmission.
– The length of the directory extension = length of the list of parameters.

  – Extension of the list: the list of parameters with their values (extension header parameters of a single object MOT, with values for the entire carousel)
  – For each object in the carousel MOT: object identifier (Transportation Id) and the contents of the object header (core + extension)
      In case of segmentation, the individual segments of the directory are sent in the data groups with Id number 6 (when the directory is not compressed) or 7 (when compressed). Thanks to this, the MOT decoder in the receiver may in the first instance integrate these segments, allowing them to reassemble and then present carousel of objects recognized in the directory. During subsequent transmissions, the MOT decoder controls the version number of each object placed in the carousel and thus also in the MOT directory. In case of detection of changes, the updates the carousel object are applied.

7.6   Broadcasting Websites

The vast information contained in the Internet prompts to use them also for the enrichment of messages in digital radio. However, the Internet protocols:

• TCP/IP (Transmission Control Protocol/Internet Protocol) to control the accuracy of the transmission on the Internet
• HTTP (Hypertext Transfer Protocol) for exchange of client information with HTTP servers

require a two-way communication, so they are not suitable for transmission directly in the DAB system. The HTTP server is essentially a collection of files, so selected applications (web) transmitted via DAB channels to a buffer in the receiver can be treated as a local server BWS (Broadcast Website), with a surfing mechanism within these applications. Browsing within such a system is based on the MOT protocol using carousel of HTTP applications and the MOT directory. Sources are identified by URLs (Uniform Resource Locators), or the URNs (Uniform Resource Names), commonly referred to as the URIs (Uniform Resource Identifiers).

Details of such mechanism are described in the Specifications [14–16] of the DAB system.

7.6.1   Organization of Transmission

Internet web pages contain links encoded as a Uniform Resource Locator (URL) not only within a single location. To allow surfing within a Broadcast Website (BWS) set, it must contain a set of linked HTML pages. In case of absence of an HTML page, the system should respond with an error message.

On transmitter site, the broadcasting system MOT BWS uses a web application server.

7.7   Overview of Messages: TopNews Carousel

In analogy to the local web surfing, the mechanism of dialing and listening to the short, compressed audio files with the latest news of different categories also can be introduced in the system DAB. For this purpose, there was introduced a specification describing the MOT carousel of TopNews objects [16]. This carousel is provided especially for the reception in a car, allowing to choose information of interest in the form of audio messages and listening when driving at any time, regardless of the current program.

7.7.1   Structure of the TopNews Applications

The TopNews services are implemented as profile in the application MOT BWS in the MOT carousel [16]. The transmission of TopNews can be realized in the PAD channel, or in the main service channel (MSC), in the packet mode. The MOT directory can be updated, but no more frequently than every 2 minutes due to a possible correction of transmission errors requiring the repetition of the directory in the next cycle of the carousel. Categories of TopNews objects and their description are determined by the service provider through the parameters of the MOT directory: "ContentSorting" and "ContentDescription". In each category, at the beginning, there must appear a verbal content index, "service index", interrupted when the browser chooses the selected object. The order of objects within the category is determined by the parameter "ContentSorting". In the receivers with screens, in parallel to the verbal information, may be displayed text description of the object indicated by the parameter "ContentDescription" in the MOT directory.

Browser provides the ability to change categories, as well as individual services. Temporary lack of an object, signaled in the index, should be marked with an audio signal.

According to the specification, the capacity of individual audio objects should not exceed 512 kilobytes, which lasts little more than 2 minutes for audio in the code of bit rate 32 kilobits per second [kbps]. It is assumed that the channel carrying that service should be limited to 64 kbps. At the receiver, the minimum buffer capacity allocated to service the TopNews in carousel demands 8 megabytes, which corresponds to the voice playback time in the code 32 kbps for more than 30 minutes.

7.7.2 Parameters of the MOT TopNews Carousel

The MOT parameters for the entire TopNews carousel include [16]:

- The type and subtype defining the general, and the specific type of objects, which implies the category and the specific type of adequate decoder on receiver site
- The profile type (ProfileSubset)
- The name (ContentName)
- The file version (UniqueBodyVersion)
- The sort of the contents (ContentSorting)
- The description of contents (ContentDescription)
- The contents of the header (Headline)
- Duration time
- The form of presentation

Individual objects of the MOT TopNews carousel are determined by the parameters of the MOT directory extension.

7.8 Electronic Program Guide EPG

Electronic program guide (EPG) described in [17, 18] aims to facilitate the selection of a specific program or service in the available up-to-date or anticipated program offer. This applies both to the currently received multiplex and other DAB multi plexes or parallel programs broadcasted in the user's location. Depending on the range of services implemented in the receiver system, the range of collected and presented information from the EPG may be different. Thus, the EPG system must operate in the receivers of various types with different capabilities of presentation, memory capacity, and the scope of the software.

Program guide in the DAB system is described in two specifications:

- ETSI TS 102818 [17], where organization of a guide system is described
- ETSI TS 102371 [18], specifying the method of encoding, compression, and guide transport

Stated below are basic information on specification of the EPG in the DAB system.

7.8.1   Specification of the EPG System

The purpose of the EPG is to equip DAB users with information about the current and near future program offer with its time parameters and attributes, which allows to use these data at the receiving end, inter alia, in order to:

- Expose programming schedule
- Navigate amid the current programs and choose the specified services to receive or record and reproduce them in the future
- Review the program offer from a variety of thematic groups

The DAB EPG is described in the XML language (Extensible Markup Language), which has been designed to describe data in a hierarchical system. According to the specification of the DAB EPG system, the EPG information is transmitted in XML files of three types:

- Information about radio stations (SI – Service Information). It is created in a file describing the division of multiplex into subchannels (radio operators), starting from appointed date.
- Information on programs (PI – Program Information), contained in files that describe the daily distribution of programs and services of designated operator ordered by start time, on a specific date, within the time 0:00:00 and 23:59:59. Description of a multiplex requires the number of objects equal to the number of subchannels multiplied by the number of days included in the EPG.
- Information about the group (GI – Group Information). It is formed in files describing thematically related services, or in series, and combined in group information within a single subchannel from the indicated date.

Any program, or service, described by its identifier or name of operator, can contain verbal brief description, version number, creation time, and a number of additional parameters that can be used to build information about the EPG in receiver.

Depending on the capacity of terminal memory reserved for the EPG decoder, specification describes two profiles: basic and advanced.

Basic profile for receivers with memory not larger than 25 kB contains a limited number of elements and attributes encoded in a simple, binary code (see appendix A in [18]).

Advanced profile also considers other attributes encoded in binary code, and further compressed.

The programs are combined in groups classified by the following types:

- Series: these programs are scheduled in certain sequences.
- Show: a collection of series.
- Concept.
- Magazine: a collection of programs edited coherently.
- Topic.
- Compilation.
- Another set.
- Another choice.

The presentation of information about the programs in the receiver depends on the manufacturer. It is expected that the information will be gradually adjusted to the individual tastes of recipients.

7.8.2   Encoding and Transport of the Electronic Program Guide in the DAB System

The EPG guide described in the XML language is distributed and updated in the DAB system as a set of multimedia MOT objects forming the carousel. The EPG guide can inform about the programs and services in multiplexes different than the currently transmitted. The EPG XML files prior to transmission are subject to simple binary coding and fragmentation. The files with the basic and advanced profiles of one multiplex must be transmitted in a single MOT carousel, because the files of advanced profile contain supplement to the data of basic profile and EPG decoder of advanced profile must decode both files.

The individual MOT object is created by each EPG XML file describing the basic profile for the specific type of information.

Additional parameters and attributes of multiplex programs, included within the advanced profile, may be further compressed and create new objects of flexible organization: the division of objects can arise from the division of channels, or days. To allow decoding together, the MOT objects containing the parameters of the basic and the advanced profiles, the files with the complementary information contain distinguishing identifiers and version numbers.

The MOT objects with the basic profile are limited. The Advanced Profile objects has no limited capacity.

To reduce the capacity of the transferred files, the data of the EPG program guide are compressed before transmission.

The broadcasted MOT objects containing the EPG information are not processed in the receivers where the feature is not implemented.

7.9   WorldDAB Digital Radio Receiver Profiles

In order to harmonize the European market for DAB receivers designed to guarantee reception of DAB and DAB+ signals, the organization *WorldDMB Forum* in collaboration with the *European Broadcasting Union* (EBU) and EICTA, after consultation with the manufacturers of integrated circuits, components, and receivers, defined the minimum requirements for DAB and DAB+ receivers [19, 20]. Receiver equipment features remain in relation to its price, hence the requirements for the different profiles of receivers. Each receiver must be able to decode the audio signal and determine parameters allowing to decode the organization of programs and data to be able to process the signal of the selected channel.

To address different channels and to determine their capacity, the service field in the DAB frame is divided into Common Interleave Frames (CIF), and every unit CIF is divided into 864 Capacity Units (CU). Each CU contains 64 bits.

The receiver functions can be expanded. It depends on enlarging the memory capacity, expansion of its software, and applying faster processors to decode data not associated with the program.

Defining different profiles of receivers allows rationalizing their selection and adjusting their types to the requirements of the national administrations and to the range of services offered in a specific area. For example, if the warning system in areas at risk of flooding is to function, receivers in this area should respond to alarm signals, what is associated with their equipment.

The terminal requirements are divided into mandatory and recommended.

Profile 1 – The Standard Radio Receiver
Spectrum:

- Mandatory reception in band III (174 to 240 MHz)

Audio:

- Mandatory decoder MPEG layer 2 (system DAB)
- Mandatory decoder MPEG-4 HE AACv2 (system DAB+)

Text:

- Mandatory displays with the name of the radio station
- Mandatory dynamic label
- Recommended the possibility of using the extended RDS character set if the display allows for it

Electronic Program Guide (EPG):

- Recommended presentation in the receivers with the appropriate display

Channel decoding:

- Mandatory decoding of minimum one sub-channel. Depending on the redundancy level (parameterized by a protection profile) it means:

   A minimum of 280 CU for audio DAB net (maximum output of MPEG 2 encoder with redundant code). It is almost one third of the main service channel's (MSC) capacity.

   A minimum of 144 CU units for maximum audio signal DAB+ with a redundant code.

FM RDS:

- Recommended reception of analog signal FM RDS

Decoding the specified services:

- Recommended reception in radios in cars:

   TPEG, TMC, information messages

- Mandatory tracking system of program from a variety of sources and switching between DAB, DAB+, and DMB, recommended also in other frequency blocks

Profile 2 – Media Radio Receiver
Spectrum:

- Mandatory reception in band III (174 to 240 MHz)

Audio:

- Mandatory decoder MPEG layer 2 (system DAB)
- Mandatory decoder MPEG-4 HE AACv2 (system DAB+)

Text:

- Mandatory displays the name of the radio station; extended dynamic text display (dynamic label+), Intellitext (text with basic viewing operations)
- Mandatory dynamic label
- Recommended the possibility of using the extended RDS character set when the display allows for it

Electronic program guide (EPG).

- Mandatory presentation
- Recommended presentation of extended profile

Channel decoding:

- Mandatory decoding minimum four subchannels. Depending on the redundancy level (parameterized by a protection profile) it means:

A minimum of 288 CU for audio DAB net (maximum output MPEG 2 encoder with redundant code). It is almost one third of the main service channel's (MSC) capacity.

A minimum of 144 CU units for maximum audio signal DAB+ with a redundant code.

FM RDS:

– Recommended reception of analog signal FM RDS

Decoding the specified services:

– Mandatory reception in radios in cars:

  TPEG, TMC, information messages

– Mandatory tracking system of program from a variety of sources and switching between DAB, DAB+, and DMB; recommended also in the other frequency blocks
– Recommended control of other frequency block programs

Minimum technical requirements for domestic and in-vehicle digital radio receivers are formulated in [21] and presented in webinars [22, 23].

References

1. ETSI EN 301 234 "Digital Audio Broadcasting (DAB); Multimedia Object Transfer (MOT) protocol"
2. ETSI TR 101 497 "Digital Audio Broadcasting (DAB); Rules of Operation for the Multimedia Object Transfer Protocol"
3. IETF RFC 2045 "Multipurpose Internet Mail Extensions (MIME) – Part 1"
4. IETF RFC 2046 "Multipurpose Internet Mail Extensions (MIME) – Part 2"
5. IETF RFC 2047 "Multipurpose Internet Mail Extensions (MIME) – Part 3"
6. IETF RFC 2048 "Multipurpose Internet Mail Extensions (MIME) – Part 4"
7. IETF RFC 2049 "Multipurpose Internet Mail Extensions (MIME) – Part 5"
8. ETSI EN 300 401 "Radio broadcast systems: Digital Audio Broadcasting (DAB) to mobile, portable and fixed receivers"
9. ETSI TS 102 367 "Digital Audio Broadcasting (DAB); Conditional access"
10. ETSI TS 101 499 "Digital Audio Broadcasting (DAB); MOT Slide Show; User Application Specification"
11. ETSI TS 102 428 "Digital Audio Broadcasting (DAB); DMB video service; User application specification"
12. ISO/IEC 13 522–5 "Information technology – Coding of multimedia and hypermedia information. Part 5: Support for base-level interactive applications"
13. ETSI TS 102 427 "Digital Audio Broadcasting (DAB); Data Broadcasting – MPEG-2 TS streaming"
14. ETSI TS 101 498–1 "Digital Audio Broadcasting (DAB); Broadcast website; Part 1: User application specification"
15. ETSI TS 101 498–2 "Digital Audio Broadcasting (DAB); Broadcast website; Part 2: Basic profile specification"

16. ETSI TS 101 498–3 "Digital Audio Broadcasting (DAB); Broadcast website; Part 3: TopNews basic profile specification"
17. ETSI TS 102 818, "Digital Audio Broadcasting (DAB); Digital Radio Mondiale (DRM); XML Specification for DAB Electronic Programme Guide (EPG)"
18. ETSI TS 102 371 V1.3.1, "Digital Audio Broadcasting (DAB); Digital Radio Mondiale (DRM); Transportation and Binary Encoding Specification for Electronic Programme Guide (EPG)"
19. 'Digital Audio Broadcasting Eureka-147. Minimum Requirements for Terrestrial DAB Transmitters', Prepared by WorldDAB, September 2001
20. 'Digital Radio Receiver Profiles', WorldDMB Forum, December 2008
21. ETSI TS 103 461 "Digital Audio Broadcasting (DAB); Domestic and in-vehicle digital radio receivers; Minimum requirements and Test specifications for technologies and products"
22. R. Lanctot, 'Car Connectivity Changes Content Consumption', DAB+ Digital Radio Technology – Implementation and Rollout, Joint WorldDAB and ABU webinar, Kuala Lumpur, Oct. 2020
23. Lindsay Cornell, 'WorldDAB updates DAB Standards', WorldDAB Technical Committee, Sept. 2020

Chapter 8
Transmission of Services in the Cumulative DAB Network: Interface STI

The DAB system is designated for broadcasting multimedia services – so providers of such services must be able to communicate in a unified way with the DAB multiplex operator. The aim of unification of such communication is to get support of the normative software for organization of the multiplex signal frames in the DAB server by multiplex operator.

Communication of the service provider with the multiplex operator has two basic functions:

- Consultation and determination of the conditions and the adequate parameters of service location in the DAB multiplex signal
- Transmission of the service itself with its parameters to the DAB server

These functions are performed via the Service Transport Interface (STI) [1] or its parts described on the three levels for implementation of only limited functions [2].

Determination of the conditions and the following parameters of service transmission requires a two-way communication between the service and multiplex operators. It is realized by the control part of the STI denoted by the acronym STI-C (Service Transport Interface-Control part).

The transmission of each component of services takes place in a unidirectional mode from the supplier or service provider to a multiplex operator through the transport interface part of the acronym STI-D (Service Transport Interface-Data part).

Transmission of services for the DAB multiplex operator can be performed in several stages in the tree configuration. The STI is then used to support call between service operator and the operator of intermediate multiplex and in the next stage between intermediate multiplexer and the DAB multiplexer.

This system of connections is illustrated in Fig. 8.1.

Each network operator collecting the multiplex services has its own unique identifier.

The slave multiplex service operator has identifier SPId (Service Provider Identifier),

© The Author(s), under exclusive license to Springer Nature Switzerland AG 2022   151
M. Oziewicz, *Digital Radio DAB+*, https://doi.org/10.1007/978-3-030-66478-7_8

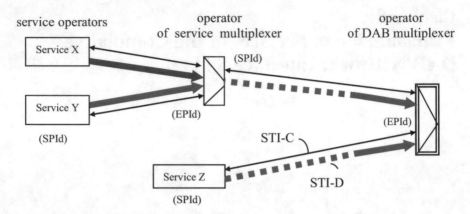

Fig. 8.1 Transmission of services in the DAB system: tree configuration

and the operator of the parent multiplex services has identifier EPId (Ensemble Provider Identifier).

For communication between the service operator and the multiplex operator, or the slave and master multiplex operators, there can be used different types of links. To decouple the organization of STI messages from the type of transmission link, the following STI layers were introduced:

- Logical layer LI (Logical Interface) of STI, for each of the parts:

  (a) The logical layer of the control part STI-C: STI-C (LI)
  (b) The logical layer of the data part STI-D: STI-D (LI)

- Adaptation layer TA (Transport Adaptation) for the control part STI-C(TA)
- The transport layer for STI-C and STI-D
- The physical layer STI(PI, X)

in accordance with Fig. 8.2.

8.1   Service Transport Interface (STI): The Logical Layer

The logical layer (LI) of STI consists of the control part STI-C(LI) and the data transmission part STI-D(LI).

8.1.1   Interface STI-C(LI)

The STI-C(LI) is realized by exchanging messages between service provider and operator of the multiplex.

Each message consists of the following fields:

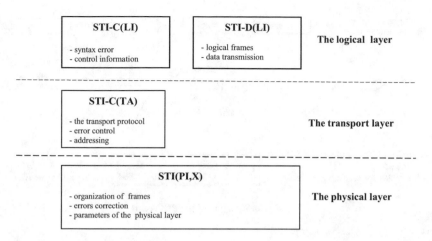

Fig. 8.2 Layers of the STI protocol

- Command code CMD (Command) identifying the message category
- Extension EXT (Extension) specifying the type of the message
- Parameter (Data fields) of the number specified by the message type
- Delimiter DELIM

The STI-C(LI) covers seven *categories of messages* corresponding to different tasks:

1. Action messages
2. Configuration messages
3. The FIG file messages
4. The grid messages FIB
5. Resource messages
6. Information messages
7. Supervision messages

The number of *types of messages* in each category ranges from 6 to 16. For specific information, the sets of messages are repeated in the form of the so-called sessions. The initiative of starting the session depends on the category of message and individual type within a category.

Ad 1) The action messages (six types) are used to inform the service operator about the takeoff time and subsequent configuration parameters. Information is transmitted by the multiplex operator at the request of service operator.

Ad 2) The configuration messages (ten types) allow to determine the configuration and position of service within the multiplex and can be grouped according to related subjects:

- Subchannel
- The type of service
- Service component (program)

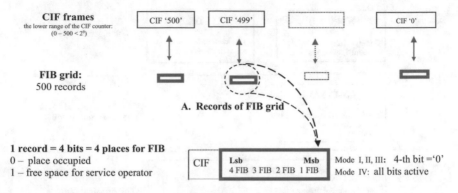

B. Construction of the FIB record

Fig. 8.3 The concept of the FIB grid

– Components of the program
– Fixed frame FIG

Defined configuration is activated by messages of action category.

Ad 3) The messages of packet FIG (eight types) are used to define, monitor, and mutually transfer sequentially selected FIG frames and indicate FIG frames to radio transmission.

This does not include the FIG frames that define the multiplex configuration (MCI): they are excluded from the package, because only the multiplex operator can them create.

Ad 4) The messages of FIB grid (six types) are used to place a synchronous switching arrangements of blocks FIB to the CIF in the structure of the signal DAB. The FIB grid consists of 500 records, each of which represents the status of a FIB block related to one CIF. The concept of grid is shown in Fig. 8.3. Status of the FIB record determines whether the content of the FIB is already defined by the multiplex operator or can be fulfilled by the service operator. Information on the FIB grid is transferred from a multiplex operator to service operator to indicate where in another 500 CIF frames there is free space to insert blocks of FIB. Each record of one grid determines the space status (free or busy).

The FIB grid is transmitted from multiplex operator to services operator in a session consisting of 10 consecutive messages, 50 records each. The FIB blocks of a service operator corresponding to the free places in the FIB grid are returned to the multiplex operator as a FIB stream through the STI-D (see next Sect. 8.1.2) synchronously with the frames of the MSC subchannels.

Ad 5) The messages of resources (including eight types)

The Resource messages are used to inform the services operator by multiplex operator about the current organization of resources of the services operator within the multiplex:

- The number of subchannel identifiers
- The number of component identifiers of the program
- The number of identifiers of services in the Fast Information Channel
- The number of components in a packet mode

Information on resources are delivered within the session. At the beginning of the session, the multiplex operator indicates the number of messages in different resource categories, which are then sent in the session.

Ad 6) The informational messages (including 16 types)

The messages in this category relate the possible and applied configurations of multiplex. They can be divided into groups of configuration messages relating to:

- The multiplex, to determine the number, or names, of possible MCI configurations to remember and currently in use
- The FIG files, to determine the number or names of the FIG files, which can be stored, as well as the number of files currently in use
- The counter of frames and time used by the service to determine the relationship between the counter of the CIF frames, the counter of the STI-D frames, and the universal time in the UTC format

Ad 7) The supervision messages (including seven types)

The categories of supervision messages allow for:

- Exchange of the messages of the error protocol of STI-C(LI)
- Transmission of the alarm status of the STI-D(LI), the physical layer interface bugs and errors apparatus

As mentioned above, communication defined by STI-C works on both sides of the link between service operator and the operator of multiplex. General algorithm of cooperation between operators during the communication session is sketched in Figs. 8.4 and 8.5. It should be considered that some messages can appear only within the session. Initiating the exchange of information is not always permitted equally to both parties.

Information obtained or verified during the session on STI-C(LI) are used to form the frames of the transport part of the STI-D(LI).

Fig. 8.4 The algorithm of creation/recognition the frames STI_C(LI)

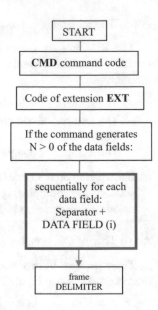

8.1.2   Logical Layer of the Service Transport Interface: The Data Part, Interface STI-D(LI)

The data part of the service transport (STI-D(LI)) defines the logical organization of a one-way transmission from the service operator (provider) to the multiplex operator. The logical interface layer forms frames comprising a single or more services in the form of data streams and streams consisting of packets of FIG or FIB files.

The frames of the STI-D(LI) have a fixed length of 24 milliseconds. Frame capacity DL (data length) is determined by the number of streams and their individual capacities ($< 2^{13}$ B).

Individual streams serve for transmission of data contributing to the main service channel (MSC) or the Fast Information Channel (FIC), in different modes. The exception is the Multiplex Service Information (MSI) data because the STI-D(LI) is not providing information about the final multiplex organization. The transport organization of the received data is decided by the multiplex operator.

The compressed audio program is transmitted in the form of frames: one MPEG frame in one frame STI-D for sampling rate 48 kHz or in the case of the 24 kHz sampling frequency the two halves of the MPEG frame in the two consecutive frames STI-D.

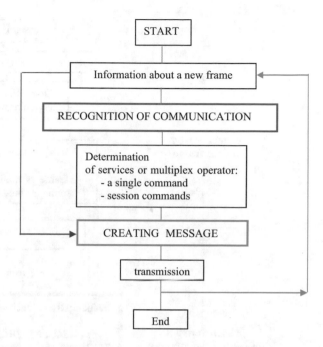

Fig. 8.5 The algorithm for determining the conditions of transmission from the service or multiplex operator through the messages STI-C(LI)

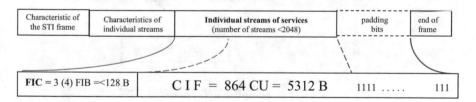

Fig. 8.6 Relation between content of the frames STI-D(LI) and CIF

Similarly for each of the other data services in the streaming mode.

Services in the packet mode are switched asynchronously: the number of packets in the frame STI-D is determined by the allocation of capacity in the CIF frame during the exchange of information between service providers and the multiplex operator via an STI-C (see previous section).

Fast Information Groups (FIG), provided for inclusion in the FIC, are transmitted in the frames STI-D in a number agreed between the service and the multiplex operators. Transmission of STI-D frames by services operator should ensure repetitions of FIG frames with the required periodicity.

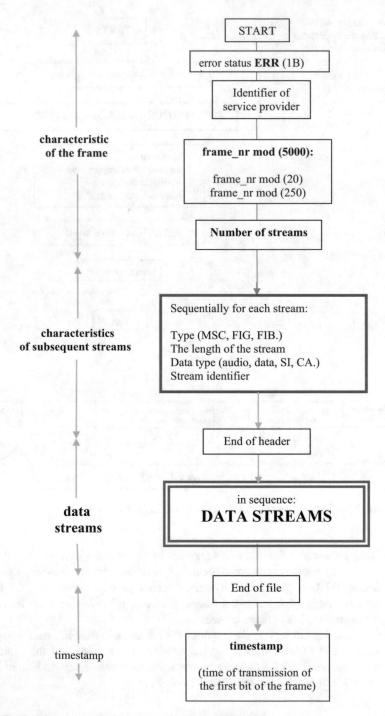

Fig. 8.7 The basic structure of the frames of protocol STI-D(LI)

Transmission of STI-D frames with a stream of Fast Information Blocks (FIB) (< 4 for the mode I) can be performed in asynchronous (with delays allowed in the server of DAB transmitter) or synchronous (no delays) incorporation into a FIC.

The STI-D frame may include simultaneously streams of different services (Fig. 8.6).

Algorithm of frame organization of the STI-D(LI) is shown in Fig. 8.7.

Alike the frame, the algorithm comprises:

A. The error status ERR (Error)
B. Characteristics of frame FC (Frame Characterization)
C. Characteristics of individual streams STC (Stream Characterization)
D. The main data stream MST (Main STream Characterization)
E. Time stamp field (optional) TIST (TIme STamp)

Ad A. *Status ERR* of the STI-D(LI) frame

The error status takes one of four values (0–3). The level zero indicates the error-free transmissions and level 3 transmission with errors disqualifying further data transmission. The content of the ERR results from the evaluation of transmission quality of frame after reception on the multiplex side.

The status ERR is determined depending on the values of the CRC independently for:

• The header (EOH)
• The individual data streams (MST)

The frames without error, or with single errors, only in the header, or only in the field of streams, are further processed. In special cases of EOH, the header of the previous frame can be taken and transmission further continued. Errors of both fields indicate the end of a transmission of a frame.

Ad B. *Frame Characterization (FC)* contains the fields specifying:

– Service Provider Identifier (SPID)
– Data Length field (DL)
– Frame number modulo 5000
– The number of individual data streams: Data Frame Count (DFCT)

Ad C. *Characteristics of individual streams (STC)* in succession for each data stream contains parameters:

Type IDentifier of transmitted data (TID). It is a code denoting, respectively:
– Subchannel MSC.
– Component of subchannel MSC.
– FIG stream in the channel FIC.
– FIB stream in the channel FIC.
– Internal information for the benefit of the operator.

- The length of the data stream STL (STream Length).
    (The length of the data stream in bytes is given in 13-bit field. Hence the length of the individual jets is in the range of 8 Kbytes.)
- Extension of the type identifier of transmitted data. This is a 3-bit field of extension code that for each type determines the classification of the content and mode of the stream:

In the MSC subchannel:

- Audio stream
- Data stream mode
- Data packet mode

In the MSC subchannel component:

- Packet mode

In the FIG stream in channel FIC:

- Service information SI
- Fast information data channel FIDC
- Information about conditional access CA

In the FIB stream in channel FIC:

- Asynchronous introduction
- Synchronous introduction
- The data STream IDentification (STID)
- For a given service provider each data stream is uniquely determined by the 12-bit code identifying the stream.

Ad D. *Individual Main Stream Data (MSD)*

In accordance with "the characteristics of the individual streams data" in the frame header, in the third part of the frame of STI-D(LI) are transported the succeeding streams.

Ad E. *The timestamp field TST (Time STamp)* optionally can be attached to the frame STI-D(LI). In this field is placed the time of transmission of the first bit in the frame STI-D(LI). The information contained in this field allows to specify the time of the frame transmission from the service provider to the DAB server. This allows synchronization of different services and service components in the organization of logical frames in the DAB multiplex.

8.2 Adaptation Layer of the STI: The Control Part

Transmission of messages of the STI-C(LI) with the help of various types of links requires the necessary information. This is done by the adaptation layer consisting of:

- Transport layer
- Network layer
- Link layer

As a result of adaptation, there are constructed frames STI-C(TA) (TA, Transport Adaptation) adapted for transport.

8.3 The Transport Layer of the Interface

The transport layer of STI consists of frames allowing the transport separately of frames STI-D(LI), frames STI-C(TA), or both frame types at the same time.

The result is a universal frame STI(PI,X) implemented in the following steps at the physical layer (PI Physical Interface) for the selected physical link X.

8.4 Physical Layer of the Interface STI

The physical layer of STI defines the projection of the frame STI(PI,X) onto packets of selected types of transmission links. Due to the different conditions of synchronous and asynchronous transmissions, this two classes of links are distinguished: synchronous and asynchronous link.

The transition from the transport layer to the network layer and detailed transformations of the frames STI(PI,X) onto the frames of different systems are described in the specification of the interface.

The extension of digital connections in the cumulative DAB network including IP was proposed in presentation [4].

References

1. ETSI EN 300 797 "Digital Audio Broadcasting (DAB); Distribution interfaces; Service Transport Interface (STI)
2. ETSI TS 101 860, "Digital Audio Broadcasting (DAB); Distribution Interfaces; Service Transport Interface (STI); STI levels"
3. ETSI TS 102 693 "Digital Audio Broadcasting (DAB); Encapsulation of DAB Interfaces", (EDI)
4. Les Sabel,'DAB+ Digital Radio", WorldDAB Workshop 2017

Chapter 9
Transmission of Multiplex Signal to Broadcasting Network: Interface ETI

As a result of forming multiplexed programs and services there is created a multiplex signal (ensemble) according with the specification [1]. After convolution encoding a multiplex frame will be transformed into a logical DAB frame. The encoding increases the volume of the frame and alike the demand for the throughput of multiplex signal, i.e., the cost of its transmission to the transmitters. The economics of distributing the signal from the output of the DAB server to transmitters of the Single Frequency Network indicates a reason to transmit multiplex signal before convolution coding (instead of a full DAB signal). Transmission of the multiplex signal can be realized through the various types of links (satellite, radio lines, public or leased telephone, etc.). To make such transmission of multiplex signal independent of the link type – the transport of multiplex was unified by the interface specification named Ensemble Transport Interface (ETI) [2].

Organization of the multiplex transport interface is separated into two layers:

(a) The part independent of the type of physical transmission network, i.e., interface ETI(NI) (Ensemble Transport Interface – Network Independent part). It allows distribution of DAB signal directly to transmitters.
(b) The interface adapted to the specific type of network ETI(NA,n) (the Ensemble Transport Interface - Network Adaptation), where n is an indicator of a type of link. The Network Adaptation layer contains additional parameters of network to be used.

Below are the indicated basic features of the ETI interface.

9.1 Interface ETI(NI)

The logical layer of the multiplex transport interface ETI is formed by the frames ETI (NI).

The single frame ETI(NI) transfers the contents of one CIF (Common Interleaved Frame) with a capacity of 864 CU (Common Unit) and supplemental information

M. Oziewicz, *Digital Radio DAB+*, https://doi.org/10.1007/978-3-030-66478-7_9

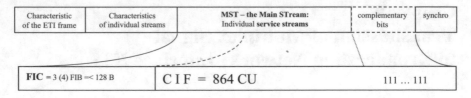

Fig. 9.1 Relation between content of the frames ETI(NI) and CIF

on organization of the channel. The CIF subchannels are transmitted in the main service stream in the ETI frame in form of data streams, one of which includes content of the FIC channel and the other are transporting subchannels of the DAB Main Service Channels (maximum 64).

The ETI frames of duration 24 ms and capacity 6144 bytes require network throughput not less than 2048 kbit/sec for this interface.

The relationship of the essential content of frames ETI(NI) and DAB frames is illustrated in Fig. 9.1.

The organization of the ETI(NI) frame is described by the algorithm in Fig. 9.2.

The ETI(NI) frame is divided into basic fields [2]:

A. The Frame Characterization, FC
B. The STream Characterization, STC
C. The End of Header, EOH
D. The Main STream data, MST
E. The End of Frame, EOF
F. The TIme STamp, TIST
G. The FRame Padding, FRPD
H. The frame Synchronization, SYNC

Ad. **A**. The Frame Characterization consists of the fields

  – *The frame count modulo 250*

The frame counter corresponds to the lower part of the CIF frame count of FIG 0/0

– *The FIC Flag. FICF*

If the FIC channel frame is arranged within the main stream, then flag FICF = 1

– *The Number of Streams, NST*

The number of streams is equal to the number of subchannels of the Main Service Channel, so NST is in the range of 1 to 64. However, when the transmitter starts – the first 15 frames may include only FIG groups due to the time interleaving of the logic frames in the Main Service Channel. Currently, the number of streams NST = 0.

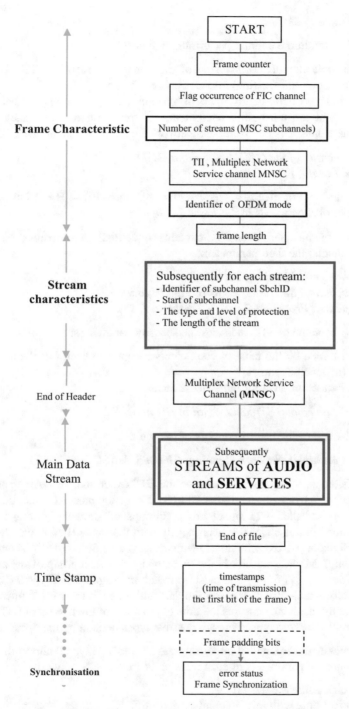

Fig. 9.2 Organization of the ETI(NI) frame

– *The Frame Phase, FP.*

This is a modulo 8 counter performing a dual role:

(a) As an indicator of the inclusion of the transmitter identifier TII (Transmitter Identification Information) into OFDM zero symbol.
(b) As a counter of consecutive pairs of bytes in the signaling channel between ETI server and the transmitter equipment, informing on time accuracy, or other message data. It is the Multiplex Network Signaling Channel, MNSC.

– *Mode Identity of the OFDM modulator, MID*
– *Frame Length, FL*

Field FL does not specify a fixed length of the frame (6144 B = 1536 words), but the number of *data words* in the frame (word = 32 b).

Ad **B**. Stream characterization STC contains sequentially for streams (i.e., the MSC subchannels) the field parameters:

– *Subchannel Identifier, SbchID*
– *Subchannel Start Address (SAD) in CU units*
– *Type and Protection Level of subchannel, TPL*

   The value of TPL indicates the security parameters:

   • Defined by the table of coefficients for the protection of the audio signal (in specification [1])
   • Protection level for multimedia data

– *The subchannel STream Length in CU units, STL*

Ad **C**. The end of the header consists of two fields:

– *The Multiplex Network Signaling Channel, MNSC*

   It is the signaling channel between the ETI server and the DAB transmitter. It consists of 2 bytes in each frame ETI(NI). Four such pairs of consecutive frames form the so-called ETI signal group. Groups of signaling messages may be transmitted in this channel synchronously with the frame (FSS – Frame Synchronous Signaling) or asynchronously with respect to the frame (ASS – Asynchronous Signaling). MNSC indicator in the frame phase defines a group of signaling bytes in the current ETI frame. The most important message FSS of type 0 contains the time information group. They set the date and time accuracy up to microseconds. Optionally, the issue concerns the time of emission of the first bit of ETI frame. In addition to the zero type, there are possible types defined by the operator.

ASS messages are divided into groups: ASS message type 0 allows the operator to determine the parameters of each transmitter in the SFN network:

• A specific address code TII
• The offset time with microsecond resolution
• 16-bit transmitter control information determined by the operator

The operator of multiplex can determine these data in FIG 0/22 group, but the network operator through the messages ASS type 0 also obtains an impact on these parameters.

– *A CRC code of the frame header*

Ad **D**. The Main STream, MST

The services of the main stream form streams of the FIC channel and the sub-channels of MSC:

- The *Stream of FIC in the FIC channel*, when the flag FICF in the characteristics of the frame is set to 1.
- The *streams of audio and services* from 1 to NST, where NST – the Number of STreams – is given in the characteristics of the frame. The length of each stream (STL) of subchannel with identity SchID is also given in the characteristics of streams in units CU.

Ad **E**. End of the Frame, EOF

Field of the end of the frame consists of two parts:

- *CRC code for the field of MST*
- *Two bytes reserved for future use*

Ad **F**. Timestamp, TST

Timestamps determine the time of maximum accuracy up to 61 ns.

Ad **G**. Frame Padding, FRPD

Since the frame has a specific capacity of 6144 bytes – the required field is complementary. It is filled with ones. In the transport layer, this field can be omitted.

Ad **H**. Sync frame, SYNC

Synchronization frame consists of two fields:

- *ERR status, Error*
- *Frame synchronization, FSYNC*

9.2 Interface ETI-NA

The frames of logical layer in interface ETI can be transmitted through digital connections with a capacity of at least 2.1 Mbit/sec. Projection of the logical structure of ETI interface onto the specific logical structure of the physical link is the content of the network adaptation of the ETI interface to specific types of links marked by acronym ETI(NA,n) (the Ensemble Transport Interface – Network Adaptation), where n is the indicator of the type of link.

Specification of interface ETI(NA,n) is associated with a specific type "n" of transmission link. Descriptions of specific links are beyond the scope of this study. Therefore, the readers interested in the specification of interface ETI on link level are referred to the original material in [2].

References

1. ETSI TS 300 799 "Digital Audio Broadcasting (DAB); Distribution interfaces; Ensemble Transport Interface (ETI)"
2. ETSI TS 102 693 "Digital Audio Broadcasting (DAB); Encapsulation of DAB Interfaces", (EDI)

Chapter 10
Conditional Access Mechanisms in the DAB System

The Conditional access in the DAB system is a set of procedures responsible for controlling access to authorized categories of recipients to the specific DAB data and services [1]. An access lock is implemented by a controlled distortion of the useful signal in a process called scrambling. Scrambling involves adding the bits of the specified information program with a string of bits from a pseudo-random generator. Reading such a signal in the receiver requires summation with identical re-synchronized pseudo-random sequence. For this purpose, it is necessary to know the activation code of a pseudo-random generator. Code is made available after verifying an authorization to receive scrambled information. Conditional access of the DAB system includes mechanisms allowing to distribute entitlements to specified (purchased) content and services.

10.1 Applications of the Conditional Access System

Restricted access to certain services of the DAB system may be due to various reasons:

- A condition of payment for certain services
- The confidentiality of certain communications and data, e.g., for the authorized state services
- The protection of minors from access to specified content
- Reservation of specified services to a closed group of users, such as chain stores
- Limiting the information to a specific geographical area

An offer of the DAB system may comprise various types of multimedia services. Self-funded system requires that at least some of them were paid, and thus not available to unauthorized persons.

M. Oziewicz, *Digital Radio DAB+*, https://doi.org/10.1007/978-3-030-66478-7_10

The mentioned functions are secured by conditional access system (CA) of the DAB radio. Conditional access system allows for flexible distribution of entitlements for:

- Operator of the CA system
- Operators of services in the DAB system
- Different groups of users of the system

The DAB specification considers the various options giving a general framework for the introduction of conditional access by entering the mechanisms of transport of the CA message system and its parameters. The DAB specification does not define a definite conditional access system, leaving in this respect the freedom of national administrations and individual operators.

10.2   Physical Implementation of the CA System

A restricted access to specific programs or services in the DAB is realized through distortion of the data in the scrambling process.

The concept of scrambling mechanisms on the transmission side and the descrambling on the receiving side is shown in Fig. 10.1.

The scrambling involves summation modulo 2 of the useful bits of service with the sequence of bits from a pseudo-random binary sequence (PRBS) generator. An example of a PRBS generator is an extended Mealy generator in accordance with the specifications of Eurocrypt [2].

Initialization of the generator follows the introduction of the Initialization Word (IW) to PRBS buffer, of a length of 10 or 11 bytes, depending on the mode of transport in a particular subchannel service. Initialization word consists of two parts:

- *Initialization modifier (IM)* computed anew with each successive DAB logic frame. At the receiver, this part of the initialization word is reproduced basing on the parameters of the signal (frame counter, the subchannel number in the MSC channel).
- *Control word (CW)*, periodically generated part of the initialization word. Control word is encrypted and transmitted to receiver in the Fast Information Channel (FIC) in the frames of the Entitlement Control Messages (ECM).

The scrambled services are loaded into the frames of the main service channel (MSC). In addition to the encrypted control word, the ECM contain conditions that must be fulfilled by a user of the system to receive the access to the services. Before decryption of the control word, the processor of an Access Control System (ACS) in the receiver verifies that the required conditions for the service access are fulfilled.

The role of the ACS processor in the receiver can be performed by plastic smart card.

When buying a card with certain entitlement privileges, one gets access to related services in the system DAB.

On the transmitter side:

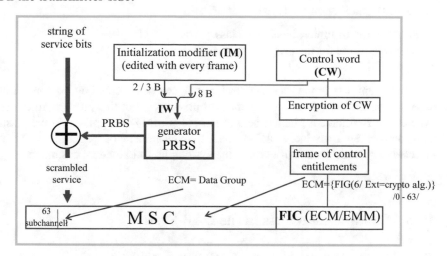

On the receiver side:

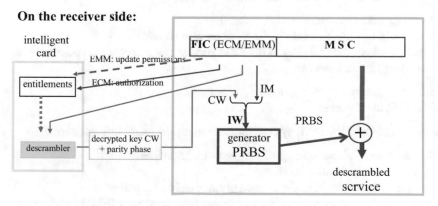

Meaning of acronyms:

PRBS - Pseudo Random Binary Sequence (generator)

IM - Initialization Modifier (word co-initiating start of PRBS generator)

CW - Control Word (generated periodically, co-initiating start of PRBS generator)

IW - Initialization Word (new edited every frame)

EMM - Entitlement Management Message

ECM - Entitlement Checking Message

Fig. 10.1 The concept of scrambling mechanism in the DAB system

The data operator may vary the range of parameters that restrict access to certain services. Depending on the type of authorization:

- Subscription of topics, level, and class
- Reservation fee for the program or service
- Time of the service

the relevant data in the access control system can be modified by the conditional access operator through the management of privileges. For this purpose, in the CA system, the Entitlement Management Messages (EMM) are applied. As the plastic cards have their own individual addresses, management of permissions can be directed to specific groups of users of the system, up to individual customers.

10.3   Conditional Access in the DAB System

The programs and services in the DAB system can be scrambled in different transport mechanisms and different components [1]:

1. Audio signal in the main service channel
2. Services in the stream mode in the main service channel
3. Data in packet mode services in the main service channel, with the possibility of scrambling at the level of:

   3.1. Data group and then split into packets
   3.2. Individual packets

The specification of the DAB system for each transport mode informs about:

- Signaling methods for encrypted programs and services
- Type of the initialization modifier
- Sources of information about applied cryptographic algorithm
- Methods of transport of entitlement management messages (EMM) and entitlement control messages (ECM)

Information about the new program or service appears in the multiplex configuration MCI in the frame FIG 0/0. Description of each new service in the FIG 0/2 contains data on the applied CA system:

- Identifier of conditional access system CAId (Conditional Access Identifier). Three bits field of identifier allows for 7 possibilities. Current specification defines the three options:

  - The lack of conditional access
  - Norwegian standard NR-MSK
  - Eurocrypt specification [2]

- for *each service component* is determined:

  - A flag of conditional access component, CA flag
    ("0" no CA, "1" CA used)

The PRBS generator is started by initiating word (IW) appropriate for the type of transmission mode.

The lower 8-byte word of initiator is a control word (CW).
The control word can be:

- Permanently fixed at both the transmitter and receiver sides

In this case, there is no need to transport the control word to the receiver.

- Generated periodically

In this case, the CW must be transmitted to the receiver. It demands the prior encryption. The cryptogram of the control word is transmitted in the ECM frame along with the encoding parameters. The 6-bit code of the type of crypto-algorithm is placed in the field of the ECM frame, besides the 1-bit phase. The phase changes with every new control word. In case of periodical generating, the control word is renewed by the module of conditional Access Control System (ACS) – e.g. smart card – every 250 frames CIF, or every 6 s (250 × 24 ms).

The DAB specification does not appoint an encryption algorithm of the control word allowing 64 possible different encryption systems.

The second part of the initialization word is an initialization modifier (IM). Construction of this part of the IW, and the method of transmission of CA-system parameters, in the ECM and EMM frames, depends on the mode of transmission of the scrambled signal and is outlined in Appendix D and described in detail in the specification of the DAB system.

10.4 Specification EUROCRYPT as a Variant of the CA System

Specification of the Eurocrypt [2] was adopted as an example of conditional access for the DAB system. The partners of this conditional access system are:

- Issuer of conditional access system deciding about the distribution and modification of authorizations to operators of programs and services
- Operators of programs and services
- Users of the system

The main tools for implementation of entitlements of the conditional access system are the so-called keys. The keys are the codes identifying service providers, or the service itself (one service provider may offer several types of services). There are different types of keys:

- *Management keys* used to load and manage entitlements
- *Operation keys* used to encrypt the transmitter control words of individual services and their decryption in receiver processor

Since each service operator holding *service key*, SK, may offer a variety of services, each service must have its own *program distribution key*, PDK, which is associated with its own set of entitlements.

Similarly, the particular service, such as meteorology, characterized by its key PDK can be transmitted by different service or program providers with the keys SK.

Key assignment is the responsibility of the operator of a conditional access system. In particular, in the framework of its tasks, the access system operator:

– Assigns keys to service providers
– Controls the keys transmitted to the conditional access processor in the receiver

On the receiver side, the keys are loaded to the security processor ACS (Access Control System). In practice, the security processor is implemented as a smart card. ACS processor contains data of:

- Key operators
- Key services
- Entitlements associated with the keys allowing for a choice of categories according to criteria:

  – Validity period
  – The remaining to use credit on smart card
  – Reception area, etc.

The operations of security processor consist of functions:

- Checking entitlements to receive desired services
- Decryption of the control word (CW) which allows to run descrambler

The entitlement to receive a specific service can be changed by the introduction of new parameters or changing the previous service parameters of the security processor ACS. This is done with the help of Entitlement Management Messages (EMM). Messages related to the services of specified operator shall provide its parameters at the address corresponding to the identifier of the service provider Program Provider Identifier (PPID).

Messages can be targeted to specific audiences by determining the user's addresses. Addresses can define:

- Individual customers through the Unique Address (UA)
- Groups of users with the Group Customer Address (GCA)

Procedures of conditional access include also control of entitlements implemented by checking the parameters of the smart card owned by the system user. If the scanner of smart card will be part of the DAB receivers, the equipment checking a particular entitlement can be realized in a program way. It can be possible also with the help of the proper software, if necessary, to change the card parameters of specified group of users by changing their permissions through the relevant signals from the transmitter. This is the mechanism for managing authorizations.

References

1. ETSI TS 102 367, "Digital Audio Broadcasting (DAB); Conditional access",
2. EN 50 094, "Access control system for the MAC/packet family: EUROCRYPT",

Chapter 11
Development of the DAB System

Telecommunication is undergoing changes related with digitization, concepts of multimedia and hypermedia, and integration of tele and computer networks. The digitization of broadcasting systems introduced these concepts also into the radio world. System DAB is a developed project offering solutions applied now also in terrestrial TV and modern iPhones.

The phases of development of the DAB system from today's perspective include:

A. Radio with sound of high quality and additional data associated with services
B. Radio highway
C. The "information on demand" for mobile reception

The first two phases were covered by the work of the European Union project EUREKA 147 DAB. Further steps are the work of the project ACTS AC054 MEMO (Multimedia Environments for Mobiles) and the joint actions of DAB and WorldDAB consortia.

Development of the system is related with the technical achievements in the field of real-time decoding of the full number of the OFDM subcarriers (FFT, Viterbi).

According to *concept A* (Fig. 11.1), system DAB appears as a digital radio broadcasting with sound quality comparable to CD quality. Value-added services are only the enhanced RDS (Radio Data System) services. This information is transmitted in the Fast Information Channel in the Fast Information Groups.

Fig. 11.1 Organizational
infrastructure of DAB
network. Stage A

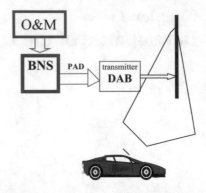

BNS   - Broadcast Network Server
O&M   - Operation and Management System
DAB   - Digital Audio Broadcasting

According to *concept B* (Fig. 11.2), system DAB introduces the advanced digital services. Classic radio programs are a key, but with the elements of multimedia offer of digital information content of interest to broad audiences. These can be information about local services, parking, traffic, and transport messages.

According to the currently implemented *concept C* (Fig. 11.3), system DAB is extended with return channel carried by mobile telephony. In this form, it is an information super-highway for a wide range of services. Depending on the type of terminal, the customer will be able to limit himself to listening to the selected radio program, or listening to the radio and select and present on display the information offered by data operators. Information for restricted audiences, or selected users of

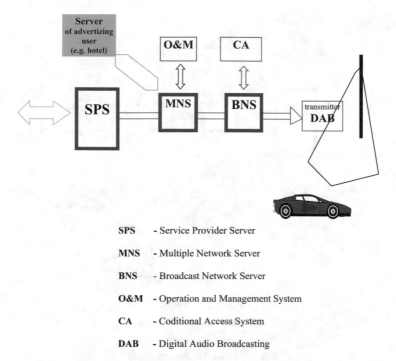

SPS   - Service Provider Server

MNS   - Multiple Network Server

BNS   - Broadcast Network Server

O&M   - Operation and Management System

CA   - Coditional Access System

DAB   - Digital Audio Broadcasting

Fig. 11.2 Organizational infrastructure of DAB network. Stage B

the system, can be ordered through mobile telephony directly from a media server of multimedia services system. Server of DAB services will be connected to the Internet and a network of selected institutions – providers of services through DAB. According to this concept, the broadcasting DAB system will be combined with the system of "information on demand" for the widest audiences, as transmitted in the cheapest way, i.e., by broadcasting.

The latest trends in the development of the system and the directions of further perspectives are discussed, among others, in works [1] and [2] and webinars by leading DAB experts in prezentations organized by EBU, WorldDAB and regional organizations in different parts of the World.

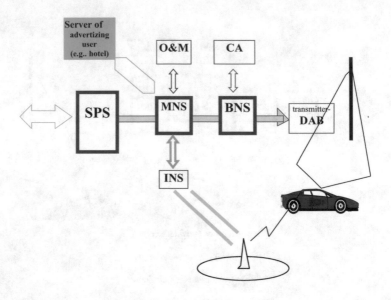

SPS	- Service Provider Server
MNS	- Multiple Network Server
BNS	- Broadcast Network Server
INS	- Interaction Network Server
O&M	- Operation and Management System
CA	- Coditional Access System
DAB	- Digital Audio Broadcasting
GSM	- Global communication Mobile System

Fig. 11.3 Organizational infrastructure of DAB network. Stage C

References

1. Lindsay Cornell, 'WorldDAB Updates DAB Standards', WorldDAB Technical Commit-tee, Sept. 2020
2. Joan Warner, 'Commercial Business Case', DAB+ Digital Radio Technology – Implemen Tation and Rollout, Joint WorldDAB and ABU Webinar, Kuala Lumpur, Oct. 2020

Appendixes

Appendix A. Basic Parameters of the DAB+ System

The carrier frequency of the transmitter is in Europe within technically reasonable ranges

$$174 - 240\ \mathbf{MHz}(\text{III VHF band})$$

with possibility of tuning every 16 kHz.

In the digital DAB or DAB+ system, the basic frequency units are blocks of size 1,536 MHz. The gross bit throughput of each block is 2309 kbit/s.

The main carrier of programs and information is the main service channel, MSC. The information in the Fast Information Channel (FIC), introduced in the transmitter simultaneously with the main channel programs, at the receiver are decoded about 384 milliseconds in advance. This is due to omission in the FIC channel of the frame time interleaving, which delays the transmission process. Adoption of constant efficiency 1/3 of convolutional encoder in this channel also accelerates the decoding. Faster transmission in FIC channel is required to forward parameters of the frame structure to the receiver to determine options of its operating system responsible for signal deconvolution and frame subchannel reconstruction.

Basic parameters of the output frame are presented in Table A.1.

Table A.1 Mode of the DAB transmitter

Parameter	Mode I
Sampling frequency	48 kHz
Frame TF	96 ms
OFDM frame gross capacity	236 544 bits
The number of symbols in OFDM frame	77
Null symbol T_0	1.297 ms
Useful symbol T_U	1 ms
Guard interval T_g	0.246 ms
Full symbol T_s	1.246 ms
Number of subcarriers	1536
Intercarrier frequency	1 kHz

Appendix B. Grouping of Convolutional Code in Channels FIC and MSC

General convolutional encoding scheme in DAB system in the individual channels is used with different perforations indices (PI).

In the audio frames, systems of codes with different perforation indices in individual fields of frames are applied.

A. Encoding Words in the Fast Information Channel (FIC)

The words outgoing the energy dispersion scrambler – prior to OFDM encoder – have a length of 768 bits (mode I). After convolutional encoder with fourfold longer mother code, the words are converted to the corresponding scheme:

$$(768 + 768 + 768 + 768 + 24 \text{ bits of the tail}),$$

and then divided into blocks:

$$(6 \cdot 128 + 6 \cdot 128 + 6 \cdot 128 + 6 \cdot 128 + 24 \text{ bits of the tail}),$$

and the sub-blocks (puncturing vectors) of 32 bits:

$$(24 \cdot 32 + 24 \cdot 32 + 24 \cdot 32 + 24 \cdot 32 + 24 \text{ bits of the tail})$$

In the Fast Information Channel (FIC), the puncturing indexes scheme in the blocks is constant with puncturing indexes PI=16, PI=15, and PI=8 for 24 bits of the tail.

The puncturing index PI $= 16$ is applied to first 84 puncturing vectors and index PI $= 15$ to the remaining 12. To the tail, the index PI $= 8$ is applied. This gives the total output codeword of length

$$84 \cdot 32 \cdot [(8 + 16)/32] + 12 \cdot 32 \cdot [(8 + 15)/32] + 24 \cdot [(8 + 8)/32] = 2304 \text{ bity}$$
$$= 3 \cdot 768 \text{ bits}$$

The code rate is therefore

$$768/(3 \cdot 768) = 1/3$$

This gives a constant code rate of third. Each frame with capacity of 256 bits because of convolutional coding is increased to 768 bits.

B. Encoding of the Applications in the Main Service Channel

In the MSC channel, where the encoder efficiency for different applications may be different, the flexibility of the encoder is increased by introducing the systems of puncturing indices.

The throughput of services in subchannels of main service channel is a multiple of 8 kbit/s (corresponding to multiple of 3 units CU per logical DAB frame).

In the subchannel of throughput $n \cdot (8 \text{ kbit/s})$, the services are transmitted in 24-milisecond logical fields in DAB frames, each of capacity $n \cdot 3CU$. These fields are elements of the encoder input.

At the output of the mother encoder, one respectively gets fields of capacity $4n \cdot 3CU$, next divided in 6n blocks of 128 bits and 24 bits of tail.

Division of 6n blocks of 128 bits into groups of L_i sub-blocks with fixed puncturing indices means that

$$\sum_i L_i = 6n \qquad\qquad (/a/)$$

Choosing in each group L_i puncturing index PI_i ($0 < PI_i < 25$) constant in the four 32-bit puncturing vectors, and in the 'tail' taking $PI = 8$, one can set up a condition that the encoder output frames have capacity equal to multiple of the units CU (64 bits):

$$\sum 4 \cdot 32 \cdot L_i \cdot [(8 + PI_i)/32] + 12 = 64 \, K \qquad\qquad (/b'/)$$

Each input frame of encoder has $3n \cdot CU$ bits, so for the output frame we have

$$K \cdot (CU) = n \cdot (3CU) + k \cdot (CU),$$

where k is the number of redundant 64-bit units CU. Hence from (b')

$$\sum L_i(8 + PI_i) + 3 = 16 \cdot (3n + k)$$

or

$$\sum L_i\, PI_i + 3 = 16 \cdot k \qquad\qquad (/b/)$$

and code rate

$$C = 3n/K = 3n/(3n + k) \qquad\qquad (/c/)$$

Assuming in /b/ maximum value of the puncturing index $PI = 24$ for all "i", we obtain for the range of k the value

$$1 <= k <= 9n \qquad\qquad (/c1/)$$

Since the numbers "n" and "k" are integers, condition /c1/ limits acceptable efficiency of C code to the ¾, 3/5, ½, 3/7, 3/8, 1/3, 3/10, 3/11, and ¼.

For a fixed subchannel throughput, several divisions for blocks of puncturing indices satisfying the conditions /a/ and /b'/ can be selected. They will ensure the code rates of the formula /c/. In order to parameterize selected divisions of blocks and puncturing indexes, the concepts of the protection profile and the protection level were introduced:

Protection profile: a fixed set of puncturing indexes $\mathbf{PI_1...,PI_k}$ for division $\mathbf{L_1,...,L_k}$
Protection level: equivalent of code rate defined for a given throughput by selection
of the protection profile

Suitable relations represent a chain:

signal throughput → frame capacity → number of sub-blocks of mother code →
→ division (L_1, ..., L_4) →protection profile (PI_1, ..., PI_4) →protection level

Protection levels from 1 (most secure) to 5 (least protection) correspond approximately to the code rates of 1/3, 3/7, ½, 3/5, and ¾. For different bit rates, these values may vary slightly.

Information about current coding parameters $\{L_i, Pi_{pi}\}$ of application are transmitted to the decoder in the receiver in a parameterized form of the protection level. For a given throughput of a subchannel, a protection level uniquely determines the puncturing parameters through specified Table [1].

For services transmitted in stream or packet mode in order to fulfill the conditions /a–c/, it is enough to split mother code into two groups of blocks (to take $k = 2$). For applications with a throughput of 8n kbit/s and protection profile $\{PI_1, PI_2\}$, the conditions on parameters take the form

$$L_1 + L_2 = 6n \qquad\qquad (/a1/)$$

$$L_1 PI_1 + L_2\,_{PI2} + 3 = 16k \qquad\qquad (/b1/)$$

$$C = 3n/(3n + k) \qquad\qquad (/c1/)$$

As mentioned above – not for each C – the condition /c1/can be satisfied.

Table B.3 Protection levels (type A) of multimedia services of bitrate 8n kbit/sec

Protection level	1A	2A	3A	4A
Code rate C	1/4	3/8	1/2	¾
Division $\{L_1, L_2\}$	{6n-3,3}	{2n-3,4n+3}	{6n-3,3}	{4n-3,2n+3}
Protection profile $\{PI_1, PI_2\}$	{24, 23}	{14, 13}	{8, 7}	{3, 2}
Inner frame capacity of audio encoder	3nCU	3nCU	3nCU	3nCU
Outer frame capacity of audio encoder	12nCU	8nCU	6nCU	4nCU

Table B.4 Protection levels (type B) of multimedia services of bitrate 8n kbit/sec

Protection level	1B	2B	3B	4B
Code rate C	4/9	4/7	4/6	4/5
Division $\{L_1, L_2\}$	{24n-3,3}	{24n-3,3}	{24n-3,3}	{24n-3,3}
Protection profile $\{PI_1, PI_2\}$	{10, 9}	{6, 5}	{4, 3}	{2, 1}
Inner frame capacity of audio encoder	12nCU	12nCU	12nCU	12nCU
Outer frame capacity of audio encoder	27nCU	21nCU	18nCU	15nCU

Assuming four protection levels for equivalent code rate C equal to ¼, 3/8, ½, and ¾, one selects appropriate protection profiles from /a1 to c1/. This task is ambiguous. The standard of the system adopted the values referred to as Type A [1] (Table B.3).

For multimedia services of bit rate equal to multiple of 32 kbit/s, or n = 4n', conditions /a–c/ can be solved for code rate 4/9, 4/7, 2/3, and 4/5. Thus, obtained additional levels of protection are referred to as type B [1] (Table B.4).

The protection level is presented in the form of a 2-bit number:

00 – level 1A (1B) (top)
01 – level 2A (2B)
10 – level 3A (3B)
11 – level 4A (4B) (lowest)

Selection of convolutional encoder parameters is the result of a compromise between:

- Encoder efficiency determined by the expected maximum error rate of transmission
- Increase of channel throughput because of redundant bits of code
- Range of time delay fluctuations contributed by the encoder

Encoding and puncturing parameters of the signal are determined by the multiplexer operator. The multiplexer operator sets the protection level for individual programs and services based on the requirements of the services operator and information on propagation conditions from the DAB network operator. Initially established average values can be assumed for the program of encoding.

Convolutional encoding increases the demanded bit rate of application. The usable throughput of the DAB system is thus limited.

Appendix C. Preliminaries on the Phasor Model
of the Multipath OFDM Signal

Information carried in the OFDM signal is included in phasors of subcarriers of the
OFDM symbols. Signal path delayed by τ deforms the phasors of the first path –
adopted as reference – and the information transmitted by the system. This modification of information following the estimated parameters of phasors can be obtained
from the phasor model.

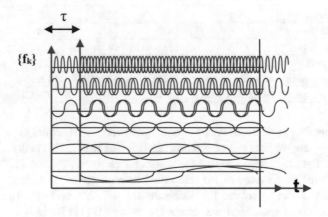

In order to pass from sinusoidal to phasor model, the phase differences between
the first and delayed by τ paths on different subcarriers are calculated:

(a) The phase difference between the first and delayed paths on subcarrier of
frequency $f_k = k/T_U$

$$2\pi \cdot f_k \cdot \tau = k \cdot 2\pi \cdot (\tau/Tu)$$

(b) The phase difference between *subsequent subcarriers* of delayed path: k and (k - 1).
In OFDM, the intertone frequency separation is equal 1/Tu, so

$$\Delta\varphi_k = 2\pi \cdot f_k \cdot \tau - 2\pi \cdot f_{(k-1)} \cdot \tau = 2\pi \cdot (\tau/Tu) = \Delta\varphi$$

This phase difference does not depend on subcarrier number.

(c) Number $k_{2\pi}$ of subchannels within full 2π turn of phasors in delayed path

$$k_{2\pi} = 2\pi/\Delta\varphi = Tu/\tau$$

(d) Frequency separation between points of full turn (in OFDM the intertone frequency separation is equal 1/Tu)

$$\Delta f = k_{2\pi} \cdot (1/Tu) = 1/\tau$$

(e) Number of full phasor turns within frequency block B

$$n = B/\Delta f = \tau \cdot B$$

The graphical presentation of phasor model is presented in Figs. 3.2, 3.3, and 3.4.

1. M. Oziewicz, 'Phasor Description of the COFDM Signal in a Multipath Channel; 6th International OFDM-Workshop (InOWo'01), Hamburg, 2001
2. M. Oziewicz, 'Phasor Description of the OFDM Signal in the SFN Network', IEEE Transactions on Broadcasting, vol. 50, no 1, pp. 63-70, March 2004

Appendix D. Transport of the Conditional Access Parameters in the DAB System

The access control is based on the following conditional statements:

- Is the program component or service a subject to limited access?
- What entitlements are necessary to gain access to its reception?
- Where are the control word parameters placed to run the PRBS generator in order to descramble data?

Since the initialization modifier IM in the control word depends on the current parameters of transmission, it is different for different types of transmission and hence different locations for the various components of audio and services in the stream transport mode, packet mode in MSC channel, and packet transport in FIC channel.

A. Signal Audio in the Main Service Channel (MSC)

Initialization modifier (IM) consists of three parts:

(a) The identifier of the MSC subchannel (Subsid), wherein the audio signal is transmitted. Subchannel identifier of the service is in the group FIG(0/4).
(b) Two complementary zero bits.
(c) The Logical Frame Counter (LFC) modulo 250 frames of FIG(0/0).

Other information of a CA system parameters is included in the Service Component Conditional Access frame (SCCA), which is also located in the FIG(0/4).

The SCCA frame contains the following information:

- What type of control word is used (CW fixed or variable)
- Whether the entitlement to conditional access has been changed (information every 32 frames)
- Whether the transmission mode of the Entitlement Control Message frame (ECM) has been changed
- The following scrambling mode
- Identifier of the frame with Entitlement Control Message (ECM) and Entitlement Management Message (EMM)
- Transport flags with information about the mode of transmission of entitlement frames of the service or its changes

Transport flags and ECM/EMM frame identifier allow for the unequivocal identification of ECM or EMM frame positioning for each component of the program or services.

ECM frame in the field of message identifier contains a 6-bit type of crypto algorithm used to encrypt the control word CW and phase of the word (bit switched along with the change of CW). Cryptogram of CW and entitlement parameters for service reception are loaded in the data field of ECM frame. The detailed organization of this field depends on the conditional access system. It is specified by the identifier CAId of the conditional access in FIG(0/2).

The EMM frame contains a flag of address type (individual or group and which), specific individual or group address of eligible customers, the type of crypto algorithm applied for the data field encryption, and in the data field the parameters of introduced or amended entitlements to receive certain services. The organization of this field is defined by the conditional access system described by the CAId.

B. Services in Stream Mode in the Main Service Channel (MSC)

The Initialization Word (IW) is built like in a point A. Information about the service subchannel and a SCCA frame of conditional access of component service is in FIG(0/4).

Acceptable transport flags of ECM/EMM frames determine their respective location.

Transport flags and identifier allow for the unequivocal positioning of ECM or EMM frames.

C. Services in the Fast Information Channel (FIC)

C1. Services scrambled before the division on the FIG 5 frames

Initialization modifier (IM) creates a 6-bit field filled with zeros and the 10-bit word of initialization modifier (IMW). The IMW word is transported in conditional access parameter FIDCCA or its extension FIDCCA_Ext (extension contains field of additional flags of transport and identification of frames ECM or EMM) of services in Fast Information Channel.

FIDCCA parameter is transported in a FIG 5 frame.

Acceptable flag transport frames ECM/EMM, respectively, define the position of its frames.

C2. Scrambling the FIG 5 frames, onto which the service is divided

Acceptable transport flags of ECM/EMM frames define respectively its positions

D. **Services in Packet Mode in MSC**

D1. Services scrambled after division onto Data Groups (DG)
The flags of transport of ECM/EMM frames define the specific variants
D2. Scrambling the packets, onto which the service is divided
Additional information about the service components transmitted in a packet
mode are in the FIG (0/3), in the field of Service Component Conditional
Access (SCCA).
Acceptable flags of transport of ECM/EMM frames define the specific variants.

Appendix E. DAB Carrier Frequencies in Blocks in the VHF Band III

Block DAB	Carrier frequency MHz
5A	174.928
5B	176.640
5C	178.352
5D	180.064
6A	181.936
6B	183.648
6C	185.360
6D	187.072
7A	188.928
7B	190.640
7C	192.352
7D	194.064
8A	195.936
8B	197.648
8C	199.360
8D	201.072
9A	202.928
9B	204.640
9C	206.352
9D	208.064
10A	209.936
10B	211.648
10C	213.360
10D	215.072
11A	216.928
11B	218.640
11C	220.352
11D	222.064
12A	223.936
12B	225.648

(continued)

Block DAB	Carrier frequency MHz
12C	227.360
12D	229.072
13A	230.748
13B	232.496
13C	234.208
13D	235.776
13A	237.448
13B	239.200

Norms and Specifications

DAB in the light of international standards and technical projects of JTC, EBU, ETSI, CENELEC, ACTS, and MEMO and specifications used in those acts.

Below the dates of publication and versions are omitted because actual are the last versions.

1. ETSI EN 300 401 "Radio broadcast systems: Digital Audio Broadcasting (DAB) to mobile, portable and fixed receivers"
2. ETSI EN 300 751 "Radio Broadcasting System; Data Radio Channel (DARC); System for Wireless Infotainment Forwarding and Teledistribution"
3. ETSI EN 300 797 "Digital Audio Broadcasting (DAB); Distribution interfaces; Service Transport Interface (STI)"
4. ETSI EN 300 798 "Digital Audio Broadcasting (DAB); Distribution interfaces; Digital baseband In-phase and Quadrature (DIQ) interface"
5. ETSI TS 300 799 "Digital Audio Broadcasting (DAB); Distribution interfaces; Ensemble Transport Interface (ETI)"
6. ETSI EN 301 234 "Digital Audio Broadcasting (DAB); Multimedia Object Transfer (MOT) protocol"
7. ETSI EN 301 700 "VHF/ FM Broadcasting; cross-referencing to simulcast DAB services by RDS-ODA 147"
8. EN 50 094 "Access control system for the MAC/packet family: EUROCRYPT"
9. EN 50 255 "Digital Audio Broadcasting system – preliminary specification of the receiver data interface (RDI)"
10. ITU-R Recommendation BS 1194 "Data Radio Channel (DARC)"
11. ITU-T Recommendation X.24 "List of definitions for interchange circuits between Data Terminal Equipment (DTE) and Data Circuit-terminating Equipment (DCE) on public data networks"
12. ITU-T Recommendation G.703 "Physical/electrical characteristics of hierarchical digital interfaces: Section 6. Interface at 2 048 kbit/sec"

© The Author(s), under exclusive license to Springer Nature Switzerland AG 2022   191
M. Oziewicz, *Digital Radio DAB+*, https://doi.org/10.1007/978-3-030-66478-7

13. ITU-T Recommendation G.704 "Synchronous frame structures used at primary and secondary hierarchical levels: Section 2.3. Basic frame structure at 2 048 kbit/sec"
14. ITU-T Recommendation G.706 "Frame alignment and cyclic redundancy check (CRC) procedures relating to basic frame structures in Recommendation G.704"
15. ITU-T Recommendation H.221 "Frame structure for a 64 to 1920 kbit/sec channel in audiovisual teleservices"
16. ITU-T Recommendation H.242 "System for establishing communication between audiovisual terminals using digital channels up to 2 Mbit/s"
17. ITU-T Recommendation H.263 "Transmission of non-telephone signals; Video coding for low bit rate communication"
18. IEC 958 "Digital audio interface"
19. ISO/IEC 646 "Information technology – ISO 7-bit coded character set for information interchange"
20. ETR 165 "Human Factors (HF); Recommendation for a tactile identifier on machine readable cards for telecommunication terminals"
21. EBU B/TPEG Project: BPN 027-1 "TPEG specification – Part 1: Introduction, Numbering and Versions TPEG-INV/004"
22. EBU B/TPEG Project BPN 027-2 "TPEG specifications – Part 2: Syntax, Semantics and Framing Structure TPEG-SSF_1.2/002"
23. ETSI ES 201 735 "Digital Audio Broadcasting (DAB); Internet Protocol (IP) datagram tunnelling"
24. ETSI ES 201 736 "Digital Audio Broadcasting (DAB); Network Independent Protocols for Interactive Services"
25. ETSI ES 201 737 "Interaction channel through Global System for Mobile communications (GSM), the Public Switched Telecommunications System (PSTN), Integrated Services Digital Network (ISDN) and Digital Enhanced Cordless Telecommunications (DECT)"
26. ETSI TS 101 428 v1.2.1 "Digital Audio Broadcasting (DAB); DMB video service; User Application Specification"
27. ETSI TR 101 495 "Digital Audio Broadcasting (DAB); Guide to DAB standards; Guidelines and Bibliography"
28. ETSI TR 101 496-1 "Digital Audio Broadcasting (DAB); Guides and rules for implementation and operation; Part 1: System outline"
29. ETSI TR 101 496-2 "Digital Audio Broadcasting (DAB); Guides and rules for implementation and operation; Part 2: System features"
30. ETSI TR 101 496-3 "Digital Audio Broadcasting (DAB); Guides and rules for implementation and operation; Part 3: Broadcast network"
31. ETSI TR 101 497 "Digital Audio Broadcasting (DAB); Rules of Operation for the Multimedia Object Transfer Protocol"
32. ETSI TS 101 498-1 "Digital Audio Broadcasting (DAB); Broadcast website; Part 1: User application specification"
33. ETSI TS 101 498-2 "Digital Audio Broadcasting (DAB); Broadcast website; Part 2: Basic profile specification"

34. ETSI TS 101 498-3 "Digital Audio Broadcasting (DAB); Broadcast website; Part 3: TopNews basic profile specification"
35. ETSI TS 101 499 "Digital Audio Broadcasting (DAB); MOT Slide Show; User Application Specification"
36. ETSI TS 101 735 "Digital Audio Broadcasting (DAB); Internet Protocol (IP) datagram tunnelling", Sophia Antipolis
37. ETSI TS 101 736 "Digital Audio Broadcasting (DAB); Network Independent Protocols for Interactive Services", Sophia-Antipolis
38. ETSI TS 101 737 "Digital Audio Broadcasting (DAB); DAB Interaction Channel through Global System for Mobile communications (GSM); the Public Switched Telecommunications System (PSTN); Integrated Services Digital Network (IDSN); and Digital Enhanced Cordless Telecommunications (DECT)", Sophia-Antipolis
39. ETSI TS 101 756 "Digital Audio Broadcasting (DAB); Registered Tables"
40. ETSI TS 101 757 "Digital Audio Broadcasting (DAB); Conformance Testing for DAB Audio"
41. ETSI TS 101 758 "Digital Audio Broadcasting (DAB); Signal strengths and receiver parameters; Targets for typical operation"
42. ETSI TS 101 759 "Digital Audio Broadcasting (DAB); Data Broadcasting – Transpa-rent Data Channel"
43. ETSI TS 101 860, "Digital Audio Broadcasting (DAB); Distribution Interfaces; Service Transport Interface (STI); STI levels"
44. ETSI TS 101 993 "Digital Audio Broadcasting (DAB); A Virtual Machine for DAB; DAB Java Specification"
45. ETSI TS 102 182 "Emergency Communications (EMTEL); Requirements for communications from authorities/organizations to individuals, groups or the general public during emergencies"
46. ETSI TS 102 367 "Digital Audio Broadcasting (DAB); Conditional access"
47. ETSI TS 102 368 "Digital Audio Broadcasting (DAB); DAB-TMC (Traffic Message Channel"
48. ETSI TS 102 371 V1.3.1,"Digital Audio Broadcasting (DAB); Digital Radio Mondiale (DRM); Transportation and Binary Encoding Specification for Electronic Programme Guide (EPG)"
49. ETSI TS 102 427 "Digital Audio Broadcasting (DAB); Data Broadcasting – MPEG-2 TS streaming"
50. ETSI TS 102 428 "Digital Audio Broadcasting (DAB); DMB video service; User application specification"
51. ETSI TS 102 563" Digital Audio Broadcasting (DAB); Transport of Advanced Audio Coding (AAC) audio"
52. ETSI TS 102 632 "Digital Audio Broadcasting (DAB); Voice Applications"
53. ETSI TS 102 652" Digital Audio Broadcasting (DAB); Intellitext; Application specification"
54. ETSI TS 102 693 "Digital Audio Broadcasting (DAB); Encapsulation of DAB Interfaces", (EDI)

55. ETSI TS 102 818,"Digital Audio Broadcasting (DAB); Digital Radio Mondiale (DRM); XML Specification for DAB Electronic Programme Guide (EPG)"
56. ETSI TS 102 978 "Digital Audio Broadcasting (DAB); IPDC Services; Transport specification"
57. ETSI TS 102 979 "Digital Audio Broadcasting (DAB); Journaline; User application specification"
58. ETSI TS 102 980 "Digital Audio Broadcasting (DAB); Dynamic Label Plus (DL Plus); Application specification"
59. ETSI TS 103 176 "Digital Audio Broadcasting (DAB); Rules of implementation; Service information features"
60. ETSI TS 103 461 "Digital Audio Broadcasting (DAB); Domestic and in-vehicle digital radio receivers; Minimum requirements and Test specifications for technologies and products"
61. ETSI TS 103 466 "Digital Audio Broadcasting (DAB); DAB audio coding (MPEG LAYER II)"
62. IETF RFC 1945 "Hypertext Transfer Protocol – HTTP/1.0"
63. IETF RFC 2068 "Hypertext Transfer Protocol – HTTP/1.1"
64. IETF RFC 1738 "Uniform Resource Locator (URL)"
65. IETF RFC 2045 "Multipurpose Internet Mail Extensions (MIME) – Part 1"
66. IETF RFC 2046 "Multipurpose Internet Mail Extensions (MIME) – Part 2"
67. IETF RFC 2047 "Multipurpose Internet Mail Extensions (MIME) – Part 3"
68. IETF RFC 2048 "Multipurpose Internet Mail Extensions (MIME) – Part 4"
69. IETF RFC 2049 "Multipurpose Internet Mail Extensions (MIME) – Part 5"
70. ISO/IEC 11172-3. "*Coding of Moving pictures and associated audio for digital storage media at up to 1.5 Mbit/s - Audio Part*". International Standard, 1993. Se encounter end la documentación complementaria.
71. ISO/IEC 13 522-5 "Information technology – Coding of multimedia and hypermedia information. Part 5: Support for base-level interactive applications"
72. ISO/IEC 13 818-3. "*Information Technology: Generic coding of Moving pictures and associated audio - Audio Part*". International Standard.
73. ISO/IEC 13 818-6 "Information technology – Generic coding of moving pictures and associated audio information. Part 6: Extension for Digital Storage Media Command and Control
74. ISO/IEC 13 818-7. "*MPEG-2 advanced audio coding, AAC*". International Standard
75. ISO/IEC 14 496-1 "Information Technology - Coding of Audio-Visual Objects—Part 1: System"
76. ISO/IEC 14 496-3 "Information Technology - Coding of Audio-Visual Objects—Part 3 – Audio"
77. ISO/IEC 23003-1 "Information Technology – MPEG audio technology – Part 1: MPEG Surround"
78. AC054 MEMO Multimedia Environment for Mobiles "Draft basic elementary service definition (Teleservices)"
79. AC054 MEMO Multimedia Environment for Mobiles," Information Package"

80. AC054 MEMO Multimedia Environments for Mobiles "Definition of Support Services"
81. AC054 MEMO Multimedia Environments for Mobiles "Final Specification of API"
82. AC054 MEMO Multimedia Environments for Mobiles "Application Software", LUTCHI,
83. AC034 and AC054 OnTheMove and MEMO "Using DAB as a Testbed for a Mobile Middleware System"
84. AC054 MEMO Multimedia for Mobiles "Implementation Guidelines for Multimedia Broadcast"
85. AC054 MEMO Multimedia for Mobiles "Protocol Standards for Broadcasting and Interaction"
86. AC054 MEMO Multimedia for Mobiles "Components for Trials"
87. AC054 MEMO Multimedia Environments for Mobiles "Project Overview"
88. ACTS Mobile Summit '98 "Toward High Integrated Terminals for a Hybrid DAB/GSM Communication System for Mobile Multimedia Services"
89. AC054 MEMO Multimedia for Mobiles "MEMO System Functional Specification"
90. AC054 MEMO Multimedia for Mobiles "Exploitation/ Market Implications of MEMO with DVB-T"
91. AC054 MEMO Multimedia for Mobiles "MEMO/ DVB-T Prototype"
92. CENELEC EN 62106, "Specification of the radio data system (RDS) for VHF/FM sound broadcasting in the frequency range from 87,5 to 108,0 MHz"
93. MEMO Specifications Version 1.1.

Protocol Specifications:
PS1. Broadcast Network Submission Protocol
PS2. Mobility Management Protocol
PS3. Extended TCP
PS4. Access Control Protocol
System Function Specifications:
SFS1. Personal Service Routing
SFS2. Personal Service Transport
SFS3. Interactive Broadcast
SFS4. Mobility Management
SFS5. Security
API Specifications
AS1. Mobile Terminal API
AS2-1. Correspondent Node API (Low Level)
AS2-2. Correspondent Node API (High Level)
System Reference Document
SRD1. MEMO System Reference Model
SRD2. MEMO System Architecture

94. Workshop on Multimedia for Mobiles M4M "Developing the Infrastructures for Integrated Broadcasting and Telecommunications Services"

95. ERTICO Committee on DAB-based Multimedia ITS Applications "DAB-based Multimedia ITS Applications. ERTICO strategy for implementation"
96. European Commission Brussels, COM (97)623 "Green Paper on the Convergence of the Telecommunications, Media and Information Technology Sectors, and the Implications for Regulation. Towards an Information Society Approach"
97. Federal Ministry of Economics and Technology 'Introduction of digital broadcasting in Germany: "Launch Scenario 2000". Status report and recommendations by the "Digital Broadcasting" Initiative on the digitisation of radio and television taking account of the cable, satellite, and terrestrial paths
98. ETSI ES 201 980 "Digital Radio Mondiale (DRM); System specification"

ETSI 300 744 "Digital Video Broadcasting (DVB): Framing structure, channel coding and modulation for digital terrestrial television"

Index

Printed in the United States
by Baker & Taylor Publisher Services